MIRACLE MEMORY METHODS

For An Amazing Memory At Any Age

By

Don McKinley

"Neither age nor intelligence is a key factor in **The Ability to Remember**!"

Journal of American Medical Association, 1999

Guess what!! Scientists now know that memory loss is not part of normal aging! You can live into your seventies, eighties, and even nineties, as sharp as a tack. You don't have to accept that memory loss is going to happen to you.

Second Printing

ABOUT THE AUTHOR
Don McKinley, CFP, CHFC

Mr. McKinley has had a varied career. A resident of Mission Viejo, California, he has 10 children and 31 grandchildren. He has a background in Accounting and Financial Planning, as well as Teaching, Acting, Singing, and as a Stage Musician. He has been teaching Principles of Memory Training for over 20 years. He has also written 9 other books on a variety of subjects.

FOREWORD
MEMORY

Donald O. Shively, Management training and memory training for over 35 years. Co-author of Speaking Without Notes and Listening, the Key to Communication.

It is my pleasure to assure you that the statement "THERE IS NO SUCH THING AS A POOR MEMORY" is no myth, it is absolutely true!

Please consider for a moment, if you place your right arm in a sling for several months, you would lose all use of it. It would take long, painful months of concentrated therapy to regain any semblance of its former usefulness.

So it is with your memory. If you have been dependent on written notes and similar devices to record the facts and information you want to save for future use, you have placed your memory in a sling. It is going to take a great deal of dedicated effort to bring it to the point of sustained usefulness, where it ought to be.

It was one of the privileges of my position as ASSOCIATE DIRECTOR of the INTERNATIONAL SCHOOL OF MEMORY TRAINING for over 15 years to select and train men and women who would teach our course through-out the Unites States.

Of the many hundreds of instructors observed by me, Don McKinley was truly outstanding. He had the innate ability to "communicate." When, for any reason, it was impossible for me to instruct one of my own classes, I always called on Don. He never failed to delight the students with the outstanding ability to teach them to improve their mental agility by improving their "memory skills."

I have been retired now for several years, but there are two areas of these methods that I use more frequently than the others. One is the instant conversion of a date to the day of the week. The other is remembering telephone numbers. I am able to look up a number, close the book, and carryon a conversation. Then days, weeks, or months later, when I have occasion to talk to that individual or firm, the mention of the name will immediately trigger the number in my mind. I use both of these methods every day to make certain that I never lose that ability. It is just as easy for me now, at age 84, as it was 70 years ago.

CONTENTS

PREFACE

> "No one has to have a poor memory! Remembering is simply the ability to recall stored information. The basic difference between a person who is able to recall this information and one who is not, lies in MEMORY TRAINING."
>
> —Don A. McKinley

"NO ONE HAS TO HAVE A POOR MEMORY"

Isn't it strange that most educators teach their students what to learn, but fail to teach them how to learn or how to remember. What would it be worth to learn how to remember or to have a retentive memory? *$1,000? $5,000?* $10,000? How much time would you be willing to invest each day to acquire that memory? Three hours? Two hours? One hour? How about thirty minutes?

Would it surprise you to know that you already have the memory, but what you lack is the "ABILITY TO RECALL." You will be amazed at your proficiency in accomplishing seemingly impossible memory feats, once you have mastered our techniques of memory training. And it is not at all difficult! But you will have to invest some of your time to learn these methods we will introduce to you.

NO, YOU DON'T HAVE TO BE A GENIUS OR EVEN SUPER SMART TO LEARN THESE SKILLS. They are available to anyone who has the desire, regardless of age. Young children as well as senior citizens can easily learn these techniques. As a matter of fact, your recall ability should be better at 50 that it is at 20, and can be, if you master the methods we will teach you.

If you have no "system" of recall, it means you have jammed information into your mind like throwing papers into a filing cabinet without any order or logic. The result would be the same. You would have a very difficult time in retrieving the information you are seeking. Make no mistake, however, the memory systems we introduce to you, must be practiced until they become a "habit." You will be excited with your newly learned skills and how they will serve you in all areas of your life.

LOSING YOUR MEMORY

DEAR DR. LAMB –I'm a 56-year- old woman in perfect health. I maintain a well-organized home and a position as head bookkeeper in a busy office. I'm by no means inactive mentally.

However, I've been noticing my memory is beginning to fail me in the most routine matters.

DEAR READER –Don't jump to conclusions. The reason many people start having problems in remembering routine things is because they're not able to concentrate. A person's concentration may decrease because of too many distractions or preoccupations.

A person's memory often will improve remarkably if he starts practicing concentration. Get a good book on "memory" from a book store. It will have a number of exercises to train your memory and concentration to lock your thoughts into your mind rather than rushing through things.

By Lawrence Lamb, M.D.

TYPES OF MEMORY

DON'T WORRY!! THERE'S A LOT ABOUT MEMORY LOSS THAT IS NORMAL

By Martha Weinman Lear

Why Do We Forget Things?

There are millions of us out here, complaining more and more about remembering less and less. Memory specialists, of whom I have interviewed, say that forgetfulness is the top health concern of people.

"Remember I asked you to remind me to call someone?" "Yes, who was it?" "I forgot."

AND "What did I go in the kitchen for?" "How do I know?" "You asked me to get it." "Get what?"

AND "1 saw whatsisname today." "Who?" "You know, whatsis-name!" "Now how would I know who that would be?" "Oh, you know, whatsisname!" "No, I don't have the faintest idea who you are talking about, where did you see him?" "It doesn't make a darn bit of difference where I saw him, I just need to know who it was!"

If you ever complain about your memory, join the crowd!!!

People worry because they do not know that this type of memory loss is normal. Yes, you heard me right, NORMAL. That means, *NORMAL*, friend. *UNIVERSAL*. So universal that the phrase "it's right on the tip of my tongue" is used in more than 40 languages.

With normal aging, what we lose is not memory in general, but rather a particular kind. You see, we have several kinds. One is *PROCEDURAL MEMORY,* which is, how-to-walk, how-to-eat, how-to-tie-a shoelace. It's what Sinatra never thought about when he sang; that

Astaire never thought about when he danced; and Tiger Woods doesn't think about when he swings a golf club. It is the memory we use unconsciously, and it is the strongest kind we have.

A second kind of memory is *SEMANTIC MEMORY,* which_covers facts. What is a key? What are eyeglasses? What is a movie?

And a third is *EPISODIC MEMORY,* which_covers experience. I've lost my keys. Where did I leave my glasses? Who was in that movie? This is the type that starts playing tag with us in the sweet fullness of time. And here is why.

That 3 pound miracle, tucked into your skull, has 100 billion neurons zapping around wildly, sending each other the electrical and chemical signals that make memories. With time, the signals weaken. Brains shrink by about half a percent a year, starting around age 30-although usually we don't notice any appreciable change for a few years after that. But here is the problem. Episodic Memory relies heavily on the front areas of the brain, the frontal lobes-the very areas that start shrinking first.

The loss isn't that great, really. It *feels* large, because we perceive a huge difference between a brain buzzing along at full strength, and one operating at, say, 95%. But it's just a slowing down. That elusive name is probably not gone-it simply takes longer to pop up. Which raises the question that everyone always asks: Is everything that ever went into my brain, still there? Answer: No one knows. (How would you find out?)

Many researchers do believe, however, that it's all there, but in altered *form.* "The disk is full," we say, and, "No room on my hard drive"-but the computer analogy is not really accurate. As Dr. Barry Gordon, a neurologist at *John* Hopkins, points out, computer memory is exact, while brain memory is fluid. Whenever you make or retrieve a memory, its pattern of signals is altered. Sort of like writing over writing. Which is why, as time passes, our memories are apt to change and deceive us.

We accept other changes in our bodies. We consider it natural that we cannot play tennis as well at 50, as we did at 20; but we cannot accept the fact that our brains also may slow down. It's simply too threatening.

Scientists who understand the whys of memory are not so easily threatened. I asked Dr. Richard E. Powers, chairman of the medical advisory board of the Alzheimer's Foundation, if he has memory problems. "A group of doctors my age were laughing about the changes we observe," he said. "At 25 we could read a scientific article once and absorb it. Now we have to read it several times. At 57, my ability to hold onto new information is not as good as it used to be-but we retain the capacity to store and use the information. It's like flypaper that's been lying on the counter *for* a long time: It's still got plenty of stick but not as much as it used to have."

We actually may be wired to forget. Consider this: If everything stuck to that mental flypaper, we would be in big trouble. We'd be overwhelmed by trivia. The longer we live, the more memories we stuff into our brains, and the harder it may become to locate any particular one. So those that we need least, the EPISODIC MEMORIES, get stored in the attic first. After all, how important is it to remember the name of the restaurant where you ate last night? What is important to remember is what "eating" means and how to eat.

MAKE YOUR MEMORY BETTER

1) Basics-See page 17
2) Get organized-See page 66
3) Associate Names-See page 72
4) Methods-Linking, see page 20
5) Letter Key, see page 30
6) Number Alphabet, see page 38
7) Number Keys, see page 53

Tests show that in absorbing new facts, we are no less competent than 20 year-olds, just slower. It will take you a little longer, *BUT YOU CAN DO IT!*

INPUT
OUTPUT

1

> "Your mind is a perfect computer, storing billions of memory energy cells. You need only to learn how to input information properly and you will have access to instant recall."
>
> Don A. McKinley

YES YOU CAN LEARN TO REMEMBER!!

If someone challenges your ability to remember, or accuses you of having a poor memory, you probably become defensive. Yet, perhaps you'd be the first to admit that you say "I forgot" or "I can't remember" several times a day. The point is, people do forget!

The reason you forget is because you have no organized way to recall facts once they have entered your mind – that is, to recall them when they are needed. Therefore, to have an effective memory you must develop a mental filing system in order to quickly recall facts when you want them.

In this respect our minds work much in the same manner as a tape recorder. When you turn on a tape recorder and push the "record" button, sounds that enter the microphone are immediately recorded on the tape. This tape may then be stored indefinitely. When this recorded information is needed, either immediately or in the future, all you must do is play back the tape. The information is readily reproduced in the exact form in which it was recorded.

And so it is with our minds. Except that your mind's "record" buttons are your FIVE SENSES. Information flows through the senses, enters your mind, and is recorded. Thus you learn.

It must be pointed out that the mind has a remarkable facility for screening information passed to it. Much of the information which enters the senses is of no value to us; thus we do not "record" it, and it is not learned. Only that information which motivates us, is interesting, makes an emotional impact, or is quickly associated with something we already know, is readily recorded as it enters our minds. All other information must be consciously learned.

By using our methods you will learn to use your mind as a mental filing system. You will receive a fact, file it away in your mind, and instantly recall it at the precise moment it is needed.

You may ask, "how will I accomplish this remarkable feat?"

How do we learn in the first place?

Memory is based on a formula that we call the four "R's."

They are simply:

RECORD + RETAIN = RECALL

THUS WE **REMEMBER** AT WILL

"RECORD" means to make a conscious effort to use your senses to visually observe information or facts which you desire to remember.

"RETAIN" means to store this information until it is needed. This is the process by which we place things in our mental filing cabinet ready for recall at will.

"RECALL" is the step needed to quickly search the mind and accurately bring forth the right fact at the right time. So, these three steps, RECORD, RETAIN, and RECALL, equal the fourth "R," which is to "REMEMBER."

THIS BOOK DEALS PRIMARILY WITH THE RECORDING AND RETAINING STEPS. THAT IS, HOW TO FILE THE MATERIAL AWAY SO THAT WHEN IT IS NEEDED YOU CAN RECALL IT QUICKLY AND EASILY.

Does this sound complicated to you? It isn't, really. Anyone can improve his memory quickly and easily with the methods taught in this book.

However, you must practice at least 15 minutes each day for several weeks in order to develop a perfect memory. Thirty minutes each day would be even better.

Then, before you know it, you will have reached your goal. You will have become a "memory expert."

Psychologists and professional memory experts have shown that one learns better by pacing himself or studying daily. Try it, and you will be amazed at how quickly you learn.

You may feel that you do not want to learn everything taught in this book because you are interested in improving just one phase of your memory. It is important that you learn all phases. Each lesson is closely related to all other lessons. You must be trained thoroughly in all the memory methods taught in this book before you can successfully specialize in a particular phase. To be a memory expert, you must learn and master each step before continuing to the next one.

2

> "The mind never thinks without a mental picture. The thinking faculty thinks of its forms in pictures. No one could ever learn or understand anything, if he had not the faculty of perception; even when he thinks speculatively, he must have some mental picture with which to think."
>
> Aristotle

THE BASICS FOR REMEMBERING

FIVE KEYS LEADING TO INSTANT RECALL:

I. OBSERVATION

"CONSCIOUS AWARENESS" Alert to things around you. Recording in your mind the things which you are consciously observing, giving attention to detail.

II. CONCENTRATION

"DIRECTED ATTENTION" Focusing your mind on one subject without being distracted by thoughts that do not pertain to that subject. Very important in the recording process.

III. ASSOCIATION

We remember best when we record new information by associating it with something we already know or remember vividly. In this way we retain that information permanently for future use. If we fail to associate properly, we will fail to recall.

IV. VISUALIZATION

Of the five senses, the sense of sight will make the strongest impression upon your mind by far, than that of any of the other senses. The pictures you are able to see in your mind are the most accurate of all mental impressions and they are the most long-lasting. This is true because the mind, like the video instant recall camera, is able to see visual images long after the original picture is removed from your external vision. We recall vivid pictures in our minds of happenings that occurred in our childhood. For example, we go to bed at night, and like a motion picture camera, see the visual image of happenings of the day or of previous days. When you physically see something, an electro impulse reaches the vision center of the brain. There is little physiological difference between the electrical signals that are activated by the physical eye and those activated by the mind's eye. This process of seeing mental images is called visualization, and the ability to utilize this characteristic of the mind and train it to your use in remembering, is the fourth key to "recall." By using these mental images in a definite systematic way, you will learn to remember easily.

V. IMAGINATION

Memory is the ability to recall at will, impressions that have been made upon our mind. If we cannot recall these mental images, then it is said, "we have forgotten". So in order to revive these impressions, we must make the images vivid and forceful so that we can enjoy instant recall. There are four basic rules for doing this:

E 1. EXAGGERATE the number of items or enlarge the objects greater than their normal size. Make them huge or gigantic. See countless numbers, literally thousands of an item.

A 2. ANIMATE the things you wish to remember as you record them in your mind. The mind is responsive to motion. Make them move, jump, dance, jiggle. Just as motion pictures are much more exciting than still pictures, it is much easier to remember something in motion than to remember a still object.

R 3. Make your mental image RIDICULOUS. See your picture doing something extremely silly, or see it dressed in a ludicrous or grotesque way. See a dog dressed like a little old lady smoking a huge cigar. A logical picture is too ordinary and does not register in the mind. Make them IMPOSSIBLE, ZANY, or ABSURD. It does take a lot of imagination to form ridiculous pictures in your mind. For children this is easier because it comes to them naturally. But for those of us who are "out of practice", it will take some exercise. However, you will see how easily it comes back to you as you utilize these memory methods.

S 4. SUBSTITUTE or picture one item in place of another. We use substitution because our minds eye cannot visualize abstract or intangible things. Therefore it becomes easier to remember by forming tangible or meaningful pictures in your mind. So we convert abstract words or other information we wish to remember into things that can be visualized.

If the pictures you imagine are Exaggerated, Animated, and Ridiculous, you will remember more vividly and recall more readily, and that, after all, is the purpose of this book. Remember – E A R S.

In the chapters that follow, you will see how important it is to understand the basics for remembering so that you can master each of the methods we will teach you. So far this doesn't sound very complicated, does it? And it really isn't. You are now ready for the first memory method we will teach you, the Linking Method.

3

> "Your memory is the most wonderful instrument in the world. You need only to know how to use it to do things that appear marvelous."
>
> --David M. Roth.

LINKING METHOD

Are you able to glance at a list of twenty objects, just once and remember everyone in sequence? Very few people could give an affirmative answer, except those who previously have studied memory training. It is almost IMPOSSIBLE for anyone to perform such a feat with an untrained memory. By the time you have finished this session, you will be able to do so quite easily.

With the LINKING METHOD you are about to learn; you will find it rather simple to memorize a list of 5, 20, 50, 100, or even more items, merely by reading or hearing the list just once.

This method is based on conscious mental association. Conscious mental association is thinking of one object in relation to the previous object on the list. This means linking objects together just as links of metal are joined together to make a chain.

Listed below are 20 items that you can memorize in a very short time by using this method:

table	house fly
tablecloth	apples
truck tire	candy bar
lady's dress	worm
two-by-four board	light bulb
watermelon	coffee cup
chair	bird
suitcase	pencil
horse	automobile
ash tray	moon

To learn this list of objects you must make vivid mental pictures. In fact, you must make ridiculous mental pictures. This may sound absurd to you, but it is a fact in memory training, that if you make a silly mental picture of the object to be remembered, the object is easier to remember. You will find that making these pictures is fun, and almost like a game.

The first item to be remembered is TABLE. Now, see a table in your mind's eye. Not just an ordinary table. Make this table ridiculous! Make it dance, make one leg shorter than the others. Do anything that makes this table different from any table you have ever seen before.

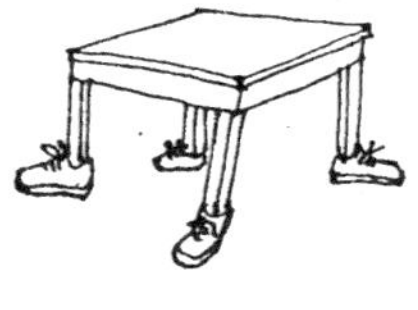

Now that you have a clear mental picture of your table, you are ready for one of the most important steps in training your memory. You are about to link together two objects with the aid of vivid mental pictures. In this case you must associate, or link, the second item on your list –TABLECLOTH --with TABLE. The normal association would be to have this tablecloth covering the table. But let's make this tablecloth do something different. Make your mental picture as absurd as possible. Stand this tablecloth on end, just so the edge touches the table top. The tablecloth is balanced and swings to music as if by magic!

Do you actually see this foolish picture in your mind's eye? It is extremely important that you do so. You must see the picture and not just the words "table" and "tablecloth." If you do not see the picture, then you may find it difficult to remember this item later.

The third item, a TRUCK TIRE, must be linked with TABLECLOTH. Not with the table, but only with the tablecloth! It is only necessary to tie each item to the object immediately preceding it. By so doing, you are forming a link in a thought chain. Once each link or item, is fastened to the one immediately preceding it, you don't have to bother with it anymore. The item will be there in your "thought-chain" when you want to recall it.

Never, under any circumstances, link three objects in the same mental picture. If you do, you might confuse the sequence of thoughts or objects. Link only two objects at a time.

Now be sure you have linked the TRUCK TIRE with the TABLECLOTH. See how dirty the tablecloth is getting from the tire. In fact, the tire is rolling across the tablecloth. See this picture clearly in your mind's eye. When you see it, you are ready to add the next link to your chain. Remember, you always link the previous object with the present one.

The next object is LADY'S DRESS, which is to be linked with the TRUCK TIRE. Out of the valve stem of the tire comes this lady's beautiful dress.

The LADY'S DRESS falls on a TWO-BY-FOUR BOARD. See the dress actually lying on top of this dirty board, what a horrible mess!

The TWO-BY-FOUR BOARD gets away from the dress and flies through the air and lands on a SLICE OF WATERMELON. In fact it squashes into the melon, causing juice to shoot out, drenching it with juice.

The WATERMELON then falls on the CHAIR.

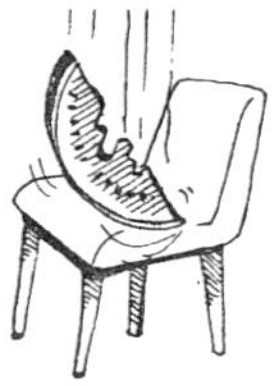

The CHAIR runs away and bangs violently into a SUITCASE.

The suitcase flies open and a midget HORSE races out.

On the HORSE'S back is a huge ASH TRAY which is teetering precariously.

The ASH TRAY finally falls off and a gigantic HOUSE FLY leaps out of it.

The HOUSE FLY is carrying a load of APPLES.

The APPLES fall and crush a CANDY BAR.

Ugh, from under the CANDY BAR crawls a WORM.

The WORM slithers across and collides with a LIGHTBULB.

The LIGHT BULB rolls away and smashes into a coffee cup.

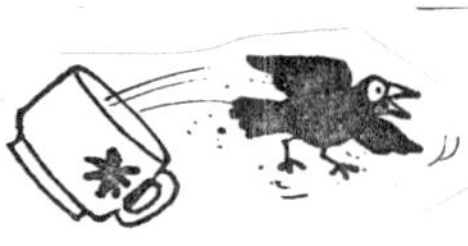

The COFFEE CUP tips over and a BIRD flies out.

The BIRD perches on a PENCIL and throws it.

The PENCIL lands on an AUTOMOBILE and starts driving it recklessly.

The AUTOMOBILE zooms away and smashes into the MOON.

The MOON falls out of the sky and lands on the TABLE you started with.

That's the entire list. If you have seen these objects in your mind's eye, you will have no difficulty in recalling them. Of course, it took much longer to explain these pictures than it does to actually see them. In fact, with practice, you can picture the item in just a fraction of a second.

Now close the book! Take a piece of paper and write down the list in the order that you learned it. If you have trouble recalling a certain item, skip it, and go on to the next item on the list. After you have "run" the entire chain, go back and see if you can fill in any links you may have missed.

If you missed one or more items in the chain, it is because you did not make a good mental picture of the object in relation to the previous object in the chain. Don't be discouraged! Before you tried this linking method, you certainly couldn't have recalled as many objects as you just did.

Now, make your own list of objects. Remember them in this same manner. You will be amazed at how quickly and easily you can do this!

Making ridiculous pictures may not be easy for you at first. Most of you have been taught to think in a logical and orderly manner. But with practice, you will find how enjoyable it is to make silly, funny pictures every time you see an object. Practice this picture-making exercise and see how easy it is.

Now you are ready to use this method in learning something you may wish to remember permanently. To show you how easy it is to use this linking method, you will now quickly learn the nine planets of our solar system in their relationship to the sun.

The nine planets are: Mercury, Venus, Earth, Mars, Jupiter, Saturn, Uranus, Neptune, and Pluto.

In making the links for these planets, we will sometimes substitute objects that are more familiar to us. This is very easy to do and works well, because all we need is a word to trigger the object we wish to remember. Here is an illustration that clearly explains how your connecting thoughts will blend together to form this chain of links. Look at it as you are reading these examples.

We start by imagining the sun bright and hot. Picture a big thermometer stuck in the SUN, and the sun's so hot that the thermometer breaks and mercury (MERCURY) squirts out of it.

Be sure you make a strong mental picture of this before you go on to the next planet.

As the mercury squirts out of the thermometer, it splashes over and hits the statue of Venus (VENUS).

Venus falls over on a large, beautiful globe of the Earth (EARTH).

On the earth there are many men from Mars (MARS) dancing and jumping.

The men from Mars climb a juniper (JUPITER) tree.

At the top of the juniper tree is an enormous piece of satin (SATURN) cloth which is flying in the breeze.

Underneath the satin it is storming and pouring down rain (URANUS).

The rain lands on King Neptune (NEPTUNE), guardian of the deep.

Neptune is chasing Mickey Mouse's dog Pluto (PLUTO).

Now close your book and write the nine planets of our solar system in the proper order. In doing so, be sure to think of the object we learned, and substitute the proper planet in its place.

Again, the four rules for **IMAGINATION** are:

1.	**EXAGGERATION**	E
2.	**ANIMATION**	A
3.	**RIDICULOUS**	R
4.	**SUBSTITUTION**	S

Try to get at least one or more of these rules into each of your pictures. Try to link just two items together at one time, thus forming a chain. Combined with visualization and association, these rules will greatly increase your ability to recall all kinds of data.

You may ask, "Now that I know how to do this silly thing, what workable use is it to me?" In later chapters you will learn other practical uses for the linking method.

For the present, you can use it for remembering shopping lists, errands to run, or to show off for your friends. Later in the book, you will use this same principle in memorizing a deck of cards in sequence, learning text material, numbers, and for many other memory applications.

The important thing for you to do now is to practice, practice practice, until linking unrelated objects together is automatic for you. When you have mastered this phase of the linking method, you will be ready to learn how to link thoughts and ideas together in a much easier learning style.

WITHOUT LINKING

4

> "The very first thing an executive must have is a fine memory. Of course, it does not follow that a man with a fine memory is necessarily a fine executive. But if he has the memory, he has the first qualification. And if he has not the memory, nothing else matters."
>
> Thomas Edison

LETTER KEY METHOD

With the LINKING METHOD you have just learned you can quickly memorize a list of objects in sequence. However, you must recall them in the same sequence in which they were memorized. You are now going to learn the LETTER KEY METHOD which will enable you to memorize items, or thoughts, in one sequence, but recall them in or out of sequence as you choose.

The Greek poet, Simonides, invented the KEY METHOD as a memory aid in 470 B.C. It was carried on by a few men through the ages, but in such a crude form that it was quite difficult for the average person to learn. David Roth, in 1918, devised a simple technique for using and teaching the KEY METHOD. Since that time, methods based on Roth's technique have been widely used, and memory training has made rapid advances.

The KEY METHOD for aiding the memory is simply a MENTAL FILING SYSTEM. By using this method you can quickly and accurately recall any fact, or series of facts, which you have "filed" in your mind. This is one of the most important memory methods because of its flexibility in helping you remember any number of things. This method can also be used to retain information permanently or for short periods of time.

If, over a period of time, you throw fact-filled scraps of paper into a filing cabinet, the cabinet soon will be filled with a mass of paper scraps. When you want a specific fact from this cluttered cabinet, you might search for hours until you find the one piece of paper with the fact you need. However, if these facts were filed in an orderly manner, and you have a key to the filing system, you can quickly find the information you want.

Most people usually "throw" information into their minds in a disorderly fashion. When they need to recall a wanted fact, they find that fact hard to "uncover" in all the clutter. If they can't find needed information in a hurry, they are apt to say, "I forgot!"

Your mind is one of the world's most efficient filing cabinets – when properly used. It is capable of holding an infinite amount of information. The KEY METHOD you are about to learn gives you the KEY to a large number of "mental drawers" in your "mental filing cabinet."

You will learn two key methods for aiding your memory – the LETTER KEY METHOD and the NUMBER KEY METHOD.

The LETTER KEY METHOD (based on the English alphabet) is shorter and therefore limited in uses. Its main value is that it can be learned quickly and put to immediate use.

The NUMBER KEY METHOD (based on numbers) which will be taught in a later chapter, takes more time to learn and has more extensive uses as a memory aid.

LETTER KEY METHOD EXPLAINED

Each Letter Key is a WORD based on the sound of the alphabet. The key has carefully been chosen to closely resemble the sound of the letter it represents. Example – A – Hay, B – Bee.

LETTER KEYS

A
HAY
B
BEE
C
SEA
D
DEAL
E
EEL
F
EFFER.
G
JEANS
H
H-BOMB
I
EYE
J
JAY
K
CAKE
L
ELBOW
M
M.D.
N
INN
INN
O
HOE
P
PEACH
Q
CUE(STICK)
R
ARMY
S
ESKIMO
T
TEE
U
HUE
V
VENUS
W
DOUBLE EWE
X
AXE
Y
WINE
Z
ZEBRA

TWENTY-SIX LETTER KEYS

A	Hay	N	Inn
B	Bee	O	Hoe
C	Sea	P	Peach
D	Deal	Q	Cue
E	Eel	R	Army
F	Effervescent	S	Eskimo
G	Jeans	T	Tee
H	H-Bomb	U	Hue
I	Eye	V	Venus
J	Jay	W	Double Ewe
K	Cake	X	Axe
L	Elbow	Y	Wine
M	M.D.	Z	Zebra

HOW TO USE KEY METHOD

As explained previously, all memory is based on mental association. The Letter Key words you have learned will serve as your permanent Key for rapid association.

Each Letter Key word represents the particular thought which is pictured on your flash card. With a little practice you can mentally trans fer the picture for each key from the card to your mind's eye. Once you have done this, you can use this picture for associating another fact or object.

A word of caution. Once you have developed a strong mental picture of each Letter Key, you should never change that picture. If you were to use a different picture each time, you would not have a fixed point on which to make your association, and you might forget your key. Don't worry about getting your keys cluttered with a lot of different things to remember. You automatically erase the last picture, or association, when you make a new one. However, if you want to retain the information, then you reinforce the association periodically until you will not forget it. Your Key can be used again in the meantime for other associations. This is a proven fact. As you practice with the Key, you will find how easy they are to use again and again.

One of the best ways to learn the Letter Key is to practice with flash cards. Each flash card having a letter, the Letter Key word, and a suggested mental picture for your Letter Key. Practice with flash cards until you can look at the letter and instantly see in your mind's eye the corresponding Letter Key. To gain full value from this memory method, it is absolutely necessary that you learn the Letter Keys to the extent that you can think with them.

Let's see how easy it will be to learn twenty-six grocery items by using the LETTER KEY METHOD. With a little practice you can learn such a list quicker with this method than you can write it down, and you don't have to worry about losing the list.

ITEM	KEY	LIST	ASSOCIATION
A	Hay	Coffee	Imagine yourself with a pitchfork pitching HAY into a steaming cup of coffee.
B	Bee	pears	See a gigantic honey BEE attacking a pear with his stinger.
C	Sea	Bread	Imagine yourself riding the waves of a stormy SEA on a loaf of bread.
D	Deal	Nuts	You are DEALING nuts instead of cards in a bridge game.
E	Eel	Beans	A huge EEL is sitting at the kitchen table shelling beans with his sharp teeth. He even has an apron on.

You make up the rest of the list and make your own association. Now, close the book and write on a sheet of paper each of the twenty-six letters and the grocery item which corresponds. Do not write the Key word. If you have missed any items, it is either because you forgot your Key word, or because you did not make a strong mental association, or picture, of the item. With a little practice you will be able to do this quickly and without missing one. The quicker you become proficient with this method the more successful you will be in training your memory. The best way to become an expert with the LETTER KEY METHOD is to use it as often as possible and in as many different ways as possible.

The LETTER KEY METHOD can be used for remembering any kind of list; or you can have your friends call out twenty-six items at random and you can recite the items back. either in order. backwards, or by skipping around. You can remember a list of errands, points of interest; and a vast number of other things.

It should be pointed out that by using this method of remembering ideas, facts, or thoughts, we usually need only one Key word to recall a phrase or even a full paragraph or more information.

Listed below are eight of the most important rules of salesmanship. These can be learned in a very few minutes with our Letter Keys. Let's learn them. Whether you're a salesman or not they are good rules to know.

ITEM	RULE OF SELLING	LIST	ASSOCIATION
A	Keep it simple	Hay	
B	Get enthusiastic	Bee	
C	Call the customer by name	Sea	
D	Show the product	Deal	
E	Keep the customer saying yes	Eel	
F	Always agree with the customer	Effervescent	
G	Ask questions to get comments	Jeans	
H	Don't knock competition	H-Bomb	

The letter keys can be used for remembering speeches, stories, and other similar applications. Suppose you were going to give a speech covering the following points:

1.Organization
2.Co-operation
3. Competition
4. Management
5. Distribution

1. Hay- see a group of men organized in cutting, piling and stacking hay.

2. Bee- see a swarm of bees drawing honey from flowers and coop-erating in placing the honey in a large bowl.

3. Sea- picture a yacht race competition in a stormy sea.

4. Deal- visualize men in a board room dealing cards to select new management.

5. Eel- imagine someone in a fish market distributing eels.

You need to make a definite picture association of the key word and the topic you are to remember. Be certain that you are able to recall these associations to your mind instantly and accurately. This becomes easier by practice. Review them a sufficient number of times until you are sure of them.

The topics considered for this example present a more difficult problem than the subjects for most speeches. This was done to illustrate the way in which difficult words can be associated with the key words. If you have sub-topics under the main thoughts, then link these with the main key word. Any idea can be reduced to a mental picture. Any two objects can always be combined in a mental image, or a thought connection formed between two ideas.

Practice is necessary to become familiar with the letter keys. You must learn to form and recall your picture associations quickly. They must instantly pop into your mind. You cannot grope for the word when you need it. The main thing is to see quickly in your minds eye the picture that represents the letter. Mental imaging requires regular practice.

By regular practice, I mean just a few minutes each day. Then these mental pictures form themselves seemingly without effort.

5

> "Your memory, like a muscle, is strengthened as you use it, and weakened if you fail to exercise it."
>
> Don A. McKinley

NUMBER ALPHABET

Do you have trouble remembering telephone numbers, or birthdays? Perhaps you have a long list of stock numbers or prices to remember. No matter what your occupation is, you have a great many numbers to recall each day. Few people can remember numbers easily. However, with the NUMBER ALPHABET, which you are about to master, you can remember any number you wish to remember, whether the number consists of one digit, twenty digits, or even more. By using the NUMBER ALPHABET you can remember an unlimited quantity of numbers.

LEARNING THE NUMBER ALPHABET

In 1648 at Marburg, Germany, Stanislaus Mink von Wenusheim gave the world the original code of letters to represent numbers. This method was improved upon by Dr. Richard Grey, M. Gregor von Feinaigle, J. R. Murden, and M. Aime Paris. In 1845, Francis Fauvel-Goraud, brought this method up-to-date, and today almost every school of memory training teaches this method.

The NUMBER ALPHABET employs the principle that numbers are difficult to learn and recall, because their meaning is abstract. Words, on the other hand, do mean something, and it is easier to think in words rather than numbers; unless those numbers are easily associated with something we already know.

For example, which address is easier to remember? A house number at 1492 Columbus Street, or one at 5604 Oak Lane? The Columbus Street address is easier to remember because we immediately associate it with the discovery of America. Unfortunately, there are few numbers as meaningful to us as 1492, or our own phone number or birthdate.

The purpose of the NUMBER ALPHABET is to make it easy for us to translate numbers into words. Then we can easily make the words meaningful, and easy to remember by association.

In order to make words from numbers, we must first use certain letters to represent the digits from one to zero.

NUMBERS TRANSLATED INTO LETTERS

The basic number to letter relationship, and the reason, is as follows:

1	is indicated by	T, d	because they both have one downstroke.
2	is indicated by	N	because it has two downstrokes.
3	is indicated by	M	because it has three downstrokes.
4	is indicated by	R	because it is the 4th letter in the Word "four."
5	is indicated by	L	because L in Roman Numerals is 50.
6	is indicated by	J	because they have the same shape.
7	is indicated by	K	because a capital K is make by a backward 7.
8	is indicated by	F, V	For no good reason.
9	is indicated by	P, b	because they have the same shape.
10	is indicated by	Z	because Zero is our last letter for the last number.

Since we have 26 letters and 10 digits we can expand the usefulness of the NUMBER ALPHABET by adding other letters with similar sounds, to the letters used in numbers 1, 6, 7, 8, 9, and O.

For number 1	we use either t, d, or th, since all three have similar sounds.
For number 6	j, soft g, sh, ch, are similar in sound.
For number 7	k, hard c, and q all sound alike. Hard g is similar.
For number 8	we use f and v, which are similar.
For number 9	p and b, are similar.
For digit 0	we use z, s, and soft c because they are similar.

You quickly will see that the numbers 1, 6, 7, 8, 9, and 0 are represented by more than one letter. In each case, the letters used have the same sound. Therefore, they have the same numerical value. By saying every letter aloud, you will notice that the mouth position in each group is the same, thus the same sound.

We now have the complete basic NUMBER ALPHABET. The vowels A, E, I, O, U and the letters W, H, and Y have no numerical value.

BASIC NUMBER ALPHABET

1	2	3	4	5	6	7	8	9	0
T	N	M	R	L	J	K	F	P	Z
D					*G	**C	V	B	S
					CH	Q			
TH					SH	**G			*C

***SOFT SOUND**

****HARD SOUND**

NO NUMERICAL VALUE

A E I O U &W H Y

Later we will explain how these vowels are used in translating numbers into words. But for now, you should thoroughly learn how to translate digits into letters, and letters back into digits.

Learn this number to letter relationship perfectly, for you will not successfully progress beyond this point until you have mastered the NUMBER ALPHABET.

EXERCISE

On the chart below, write in the numbers represented by the following letters:

1.	T ________	11.	D ________	21.	Soft G________
2.	R ________	12.	SH ________	22.	Hard C ________
3.	P ________	13.	V ________	23.	CH ________
4.	CH ________	14.	Z ________	24.	R ________
5.	B ________	15.	K ________	25.	Soft C ________
6.	Q ________	16.	M ________	26.	V ________
7.	J ________	17.	F ________	27.	Hard G ________
8.	N ________	18.	Q ________	28.	Hard C ________
9.	L ________	19.	B ________	29.	Soft G ________
10.	S ________	20.	SH ________	30.	Soft C ________

Write the letters representing the following numbers:

1.	3 ________________	11.	6 ________________
2.	8 ________________	12.	1 ________________
3.	4 ________________	13.	9 ________________
4.	6 ________________	14.	7 ________________
5.	1 ________________	15.	0 ________________
6.	9 ________________	16.	4 ________________
7.	7 ________________	17.	6 ________________
8.	5 ________________	18.	8 ________________
9.	0 ________________	19.	7 ________________
10.	2 ________________	20.	0 ________________

Be sure to do the above exercises before going on, and make sure you have them correct.

It is extremely important that you think sound, and not spelling, while learning to use the NUMBER ALPHABET. In many words the spelling and sound of the word do not correspond. You will see examples of this later in this section.

The NUMBER ALPHABET is used for remembering all kinds of numbers, and is the basis for another memory method you will learn later. It is imperative that you learn the NUMBER ALPHABET well!

The purpose of the NUMBER ALPHABET is to give number values to words. An easy way to learn it is to practice translating words into numbers. For example: The word DESK is translated 107.

Practice translating the following words into numbers:

EXERCISE

1. Saw ____________________
2. Pen ____________________
3. Table ____________________
4. Paper ____________________
5. Mind ____________________
6. Wrist ____________________
7. Lamp ____________________
8. Larva ____________________
9. Horse ____________________
10. Card ____________________
11. Speed ____________________
12. Josh ____________________
13. Dusk ____________________
14. Flat ____________________
15. Police ____________________
16. Cigar ____________________
17. Actor ____________________
18. Quiz ____________________
19. Gauge ____________________
20. Cinch ____________________

So far you have discovered how easy it is to learn the NUMBER ALPHABET, and how simple it is to translate words into numbers. In order to make this NUMBER ALPHABET more useful, we must translate numbers into words quickly. Remember, in translating numbers into words, we translate by sound only, and not by spelling.

Three important rules to remember in using the NUMBER ALPHABET:

1. WHEN LETTERS ARE SILENT THEY HAVE NO NUMERICAL VALUE.

EXAMPLE: The letter "t" in the word "often" is silent, and has no numerical value. Thus, "often" is translated to the number 82 and not 812.

2 A DOUBLE LETTER USUALLY HAS THE SAME VALUE AS A SINGLE LETTER. Remember - sound not spelling.

EXAMPLE: Jolly is translated 65 and not 655.

3. THE LETTERS W, H, X AND Y, AND VOWELS A, E, I, 0, U, HAVE NO NUMERICAL VALUE. They are used with the consonants in the Number Alphabet to form words.

EXAMPLE: The numbers 4 and 7 are translated into "r" and "k," adding vowels to make a word, we have "rake."

The examples below show how numbers are translated by sounds rather than by the way the word is spelled. These examples may look difficult to you, but don't let them scare you. There is no need to learn them by heart for they are shown to aid you in thinking in sounds rather than in spelling. The "six" group has only four sounds. They are soft G, J, sh, and ch – all soft sounds. The "seven" group has the hard sounds of K, Q, hard C and hard G. After you have studied the list below, you will find that it is not as hard as it might seem.

SPELLING VS. SOUND

T is 6 as pronounced in picture, notion, and essential
Z is 6 as pronounced in azure and seizure
S is 6 as pronounced in usual and erasure
SS is 6 as pronounced in tissue and mission

C is 6 as pronounced in ocean, delicious, and crucial
D is 6 as pronounced in soldier
SCH is 6 as pronounced in schnitzel and schnauzer
DG is 6 because D is silent in front of soft G as in ledge
TCH is 6 because T is silent in front of CH as in match
CH is 7 as pronounced in ache and chorus
CK is 7 because C is silent in front of K
PH is 8 because it always is pronounced as an F
GH is 8 as pronounced in tough and enough
X is 0 as pronounced in Xenon or Xenia
X is 70 as pronounced in fix, exact, and exert
X is 76 as pronounced in anxious
The vowel N hard G sound is a 27

In the following exercise, translate the words into numbers. To help you do this I have spelled the word correctly, and have given the correct dictionary pronunciation. As you can see, many words are not spelled the way they are pronounced.

EXERCISE
TRANSLATE WORDS INTO NUMBERS

Only translate the first three digits in each word. This is always done because in practical use we use only three digits for each word.

WORD (SOUND)	NUMBER	WORD (SOUND)	NUMBER
match (mach)	________	through (thru)	________
buffet (bu-fa)	________	palm (pam)	________
buffet (buf-et)	________	Psalms (sams)	________
cancer (kanser)	________	angel (an-jel)	________
seesaws (se-sas)	________	angle (ang-l)	________
fire wall (fyr wal)	________	subpoena (su-pe-na)	________
lodging (loj-ing)	________	unknown (un-non)	________
bobby-pin (bob-y-pin)	________	talker (tok-er)	________

WORD (SOUND)	NUMBER	WORD (SOUND)	NUMBER
pneumatic (nu mat-ik)	________	axis (ak-sis)	________
xylophone (zi-lo-fon)	________	roughage (ruf-ij)	________
St. Louis (sant lu-is)	________	weigh (wa)	________
soapbox (sop-boks)	________	X-mas (kris-mas)	________
magician (ma-jish-an)	________	film (film)	________
excellent (ek-se-lent)	________	X-ray (eks-ray)	________
nature (na-cher)	________	anchor (ang-ker)	________
depot (de-po)	________	question (kwes-chun)	________
jacket (jak-et)	________	luscious (lush-us)	________
circus (sur-kus)	________	social (so-shal)	________
amphitheatre (am-fi-the-ater)	________	mnemonist (ne-mon-ist	________
bookkeeper (book-kep-er)	________	unnecessary (un-nes-e-ser-y)	________

Now that you have mastered the NUMBER ALPHABET, and are familiar with its translation by sounds, you are ready to learn how to translate numbers into words. You must be able to do this in order to effectively use the NUMBER ALPHABET method of memory.

To translate numbers into words, you simply translate the number into its basic sound or consonant. Take the number 43 for example. The four translates into "r" and the three into "m." Now that we have the number translated into letters, we must add enough additional vowels (see rule 3 above) to make a word that means something to us. Do this by starting at the beginning of the vowels, and go through them until you hit on a vowel and consonant combination that makes a word. In this situation, the first vowel, A, fits perfectly. The word becomes RAM, which is 43. Other words for 43 are: RUM, ROOM, RIM, RHYME, HAREM, ARMY, WORM, etc.

Now let's try a larger number. The number 210 is a good example.

2	1	0
N	T	Z
	D	S
	Th	soft C

By adding vowels or W, H, or Y in various combinations, we discover a number of words. To list a few: NUTS, NIGHTS, NOTES, NETS, NOTICE, NEEDS, NEATEST, NAUGHTIEST, NUTTIEST, and numerous others.

Of course, we can put the no value letters in front of the 2 and get such words as: KNIGHTS, GNATS, HINTS, WINDS, WOUNDS, HOUNDS, HAUNTS, ENDS, HANDS, WANTS, KNOTS, WINDOWS, INDUSTRY, ENTICE, INDESTRUCTIBLE, INDISPENSABLE, and many.others.

One of the values of the NUMBER ALPHABET method is that we can pick the word that best fits the situation.

Below is an exercise to give you practice in translating numbers into words. Remember, use ONLY the first THREE numbers in each word. You may use words with more than three number alphabet sounds in them, but any sounds past the first three are not translated, as we explained earlier in this section.

In doing the following exercise on making words into numbers, use a piece of scratch paper. Write down the consonants under the number to be translated, as shown in the example above. Write in the various vowels until you have made a word. You will be surprised at how quickly you can work this exercise in this manner.

EXERCISE IN MAKING WORDS TO REMEMBER NUMBERS

Write at least 6 words for the following numbers:

01 ______________________________

04 ______________________________

09 ______________________________

13 ______________________________

16 ______________________________

19 ______________________________

22 ______________________________

25 ______________________________

28 ______________________________

33 ______________________________

37 ______________________________

39 ______________________________

40 ______________________________

41 ______________________________

54 ______________________________

56 ______________________________

57 ______________________________

61 ______________________________

66 ______________________________

68 ______________________________

72 ______________________________

74 ______________________________

78 ______________________________

81 ______________________________

82 ______________________________

89 ______________________________

90 ______________________________

92 ______________________________

96 ______________________________

Write at least 5 words for the following numbers:

010 ____________________

029 ____________________

035 ____________________

051 ____________________

079 ____________________

100 ____________________

112 ____________________

134 ____________________

204 ____________________

270 ____________________

341 ____________________

394 ____________________

409 ____________________

432 ____________________

485 ____________________

512 ____________________

575 ____________________

601 ____________________

641 ____________________

694 ____________________

700 ____________________

707 ____________________

754 ____________________

820 ____________________

841 ____________________

853 ____________________

894 ____________________

921 ____________________

940 ____________________

995 ____________________

You should have fun practicing making words out of numbers as you see numbers printed on signs, license plates, addresses, phone numbers, or wherever you find numbers.

SOME PRACTICAL USES OF THE NUMBER ALPHABET

Here are a few examples of how the NUMBER ALPHABET can be used to aid your memory. These examples were chosen for practice only. You may not wish to remember them permanently.

EXERCISE

BONES OF THE ADULT BODY		
Head	29	______________
Vertebral Column	26	______________
Ribs and Sternum	25	______________
Upper Extremities	64	______________
Lower Extremities	62	______________
Total	206	______________

HISTORY DATES		
Confucius Born	550 BC	______________
King Arthur Killed	537	______________
Magna Carta Signed	1215	______________
Gutenberg Bible	1456	______________
Pilgrims Landed	1620	______________
Boston Tea Party	1773	______________
Donut Invented	1847	______________
Dewey Decimal System	1873	______________
Pearl Harbor	1941	______________

GEOGRAPHY DATA

Death Valley	-282 ft.	______
Hoover Dam	726 ft.	______
Yosemite Falls	1,750 ft.	______
Miss-Mo. River	3,892 mi.	______
Marianas Trench	35,948 ft.	______
Mt. Everest	29,028 ft.	______
Mt. Whitney	14,495 ft.	______
California Area	158,693 sq. mi.	______

SCIENTIFIC FACTS

1 cubic ft. of water weighs 62.5 lbs.	______
Water boils at 212° F.	______
Earth orbital velocity 18.5 MPS	______
Gold weighs 19.25 times water	______
Salt weighs 2.130 time water	______
Absolute zero is -459.6	______
1 Nautical mile equals 6076.1 ft.	______
1 Meter equals 39.371 inches	______
pi is 3.14159265	______

MISCELLANEOUS FACTS

- THE EMPIRE STATE BUILDING is 1,248 feet high and weights 600 million pounds.
- ROBERT FOSTER held his breath 13 minutes, 42.5 seconds in March, 1959.
- DON GRATE threw a baseball 445 feet, 1 inch in August, 1956.
- VENUS is 67 million miles from the Sun and has a 224 day year.

REMEMBERING TELEPHONE NUMBERS			
CITY	AREA CODE	WORD	ASSOCIATION
Dallas, TX	214	Hunter	____________
Las Vegas, NV	702	Casino	____________
Salt Lake City, UT	801	Faucet	____________
Washington, DC	202	Anacin	____________
Los Angeles, CA	213	Anatomy	____________
Miami, FL	305	Muscle	____________

Now tie in the area code with four or five words representing the seven digit phone number.

Example:

619-298-1538

JET HOOP – NOW BEHAVE – DIAL MY PHONE

Now for practice, let's try some without the area code!

____________ ____________

____________ ____________

____________ ____________

____________ ____________

____________ ____________

____________ ____________

6

> "The greatest waste in the world today is not the depletion of natural resources, nor is it the devastations of war. Rather it is that man uses less than 20% of his mental ability in his lifetime. The mind, therefore, has unlimited potential and we must find a way to release it."
>
> Don A. McKinley

NUMBER KEY METHOD

The NUMBER KEY METHOD differs from the LETTER KEY METHOD in that we have a key word for each one and two-digit number, rather than for just the letters of the alphabet.

The advantages of the Number Keys are many. In the first place, you will have over one-hundred keys which you can use for instant association in remembering a long list of related lists at the same time. You can remember appointments, dates, outlines, telephone numbers, and virtually anything requiring numbers. Since most things we are required to learn, or remember, are numbered rather than lettered, the NUMBER KEY METHOD is especially valuable to us.

By linking the Number Keys together, we can learn extremely long numbers almost instantly, or a deck of playing cards in sequence in just a matter of minutes. The entire contents of a magazine or textbook can be memorized, page by page. Later in the book you will learn how to do these feats of memory, plus a number of other important uses of the Number Keys. But first, let's learn the NUMBER KEY METHOD. Remember, it is vitally important to you and your memory training, so LEARN THE NUMBER KEYS WELL.

BUILDING THE NUMBER KEYS

Since the Number Keys are based on the NUMBER ALPHABET, they are made exactly in the same manner in which you translated numbers into words. However, to expedite learning, we have made the entire list of Number Keys for you. Our research has shown that these words make better key words than others.

The key for the numeral 1 is hat. Hat was chosen from a list of over two-hundred words because we wanted all numbers from 1 to 9 to be represented by a word beginning with the letter "H." The first number, 1, in the NUMBER ALPHABET, is "t" or "d." Now trying the first vowel "a" with the letter "h" and "t" we get the word "hat." We asked ourselves if hat is a word that-can be easily associated with other things, and, of course the answer is yes.

Our word for "2" is "hen;" Han did not make a meaningful word.

This same procedure was followed for the entire list of number keys. We were very careful not to select a word that could be confused with another Key word in any way. You will find that each division: 1-9, 10-19,20-29, etc,; starts with the same letter. This was done to simplify learning.

We strongly urge you to use the key words listed here. However, if any of these Number Keys are not meaningful to you, then substitute another word which has meaning for you. Make absolutely sure, however, that this word cannot be confused with another key word.

You can learn the Number Keys easily at the rate of 40 per week. Use your Number Key flash cards in the same manner used for learning the Letter Key words. Use every Key Word at least twice each week to memorize lists of objects. By using the key words in this manner, you will quickly become proficient in their use.

NUMBER KEYS

			1	2	3	4	5	6	7	8	9	
H			HAT	HEN	HAM	HARE	HALL	HASH	HOOK	HOOF	HOOP	
		10	11	12	13	14	15	16	17	18	19	
T		TOES	TOT	TON	TIME	TAR	TAIL	TISSUE	TACK	TAFFY	TUB	
		20	21	22	23	24	25	26	27	28	29	
N		NOOSE	NET	NUN	NAME	NERO	NAIL	NOTCH	NOOK	KNIFE	NAP	
		30	31	32	33	34	35	36	37	38	39	
M		MESS	MAT	MOON	MUMMY	MARE	MAIL	MATCH	MIKE	MUFF	MOP	
		40	41	42	43	44	45	46	47	48	49	
R		RACE	RAT	RAIN	RAM	ROWER	RAIL	ROACH	RAKE	ROOF	ROPE	
		50	51	52	53	54	55	56	57	58	59	
L		LACE	LIGHT	LION	LIME	LAIR	LILY	LASH	LOCK	LEAF	LIP	
		60	61	62	63	64	65	66	67	68	69	
J		JUICE	JET	JUNE	JAM	JAR	JAIL	JUDGE	JACK	JAVA	JEEP	
		70	71	72	73	74	75	76	77	78	79	
C		CASE	CAT	CANE	COMB	CAR	COIL	CASH	COKE	CAFE	COP	
		80	81	82	83	84	85	86	87	88	89	
F		FUSE	FIGHT	FAN	FOAM	FIRE	FILE	FISH	FAKE	FIFE	FOB	
		90	91	92	93	94	95	96	97	98	99	100
P		PASS	POT	PIN	PALM	PIER	POOL	POUCH	PEAK	PUFF	PIPE	DAISIES
	0	00	01	02	03	04	05	06	07	08	09	
S	SAW	SAUCE	SEAT	SUN	SEAM	SOAR	SAIL	SASH	SACK	SAFE	SOAP	
			100	200	300	400	500	600	700	800	900	1,000
W			WET	WIN	WHAM	WIRE	WELL	WITCH	WICK	WAVE	WEB	WOODS

0
S sC Z
1
T D Th
2
N
3
M
4
R
00 SAUCE
SAUCE
0 SAW
1 HAT
2 HEN
3 HAM
4 HARE
10 TOES
11 TOT
12 TON
1000 POUNDS
1000 POUNDS
13 TIME
14 TAR
20 NOOSE
21 NET
A
22 NUN
2
23 NAME
BOB
3
24 NERO
4
30 MESS
10
31 MAT
WELCOME
J
32 MOON
Q
33 MUMMY
K
34 MARE
40 RACE
41 RAT
A
42 RAIN
2
43 RAM
3
44 ROWER
4
50 LACE
10
51 LIGHT
J
52 LION
Q
53 LIME
K
54 LAIR

5 L	6 J sG Sh Ch	7 K hC Q hG	8 F V	9 P B
5 HALL 	6 HASH 	7 HOOK 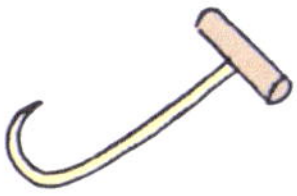	8 HOOF 	9 HOOP 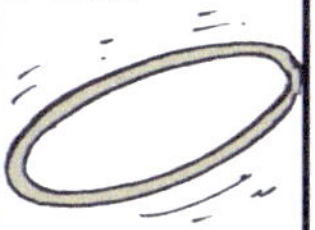
15 TAIL 	16 TISSUE 	17 TACK 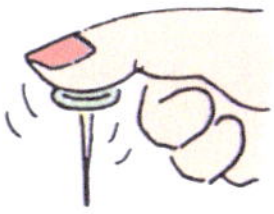	18 TAFFY 	19 TAPE 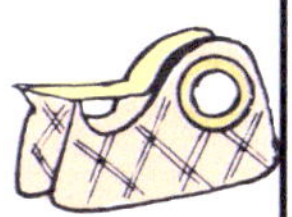
25 NAIL  5♣	26 NOTCH  6♣	27 NOOK 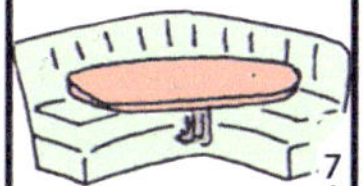 7♣	28 KNIFE  8♣	29 NAP  9♣
35 MAIL 	36 MATCH 	37 MIKE 	38 MUFF 	39 MOP 
45 RULE  5♦	46 ROACH  6♦	47 RAKE  7♦	48 ROOF 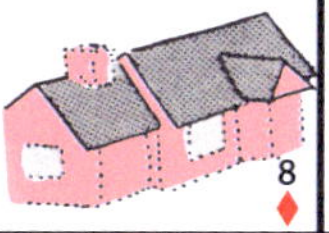 8♦	49 ROPE  9♦
55 LILY 	56 LASH 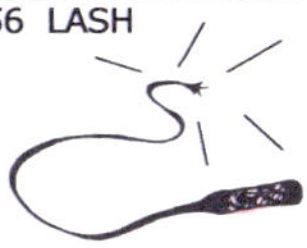	57 LOCK 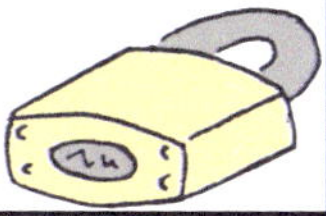	58 LEAF 	59 LIP 

0 S sC Z	1 T D Th	2 N	3 M	4 R
60 JUICE 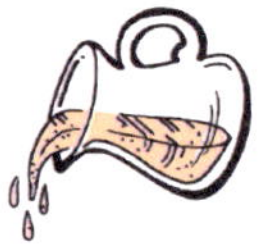	61 JET  A♥	62 JUNE  2♥	63 JAM  3♥	64 JAR  4♥
70 CASE  10♥	71 CAT  J♥	72 CANE 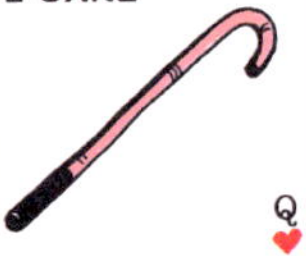 Q♥	73 COMB  K♥	74 CAR 
80 FUSE 	81 FIGHT  A♠	82 FAN  2♠	83 FOAM  3♠	84 FIRE  4♠
90 PASS  10♠	91 POT  J♠	92 PIN  Q♠	93 PALM  K♠	94 PIER 
100 DAISIES 	01 SEAT 	02 SUN 	03 SEAM 	04 SOAR 
1000 WOODS 	100 WET 	200 WIN 	300 WHAM 	400 WIRE 

5 L
6 J sG Sh Ch
7 K hC Q hG
8 F V
9 P B
65 JAIL
66 JUDGE
67 JACK
68 JAVA
69JEEP
75 COIL
76 CASH
77 COKE
Coke
78 CAFE
79 COP
85 FILE
86 FISH
87 FAKE
88 FIFE
89 FOB
25 YEARS SERVICE
95 POOL
9 FT.
96 POUCH
97 PEAK
9 PUFF
POWDER PUFF
99 PIPE
05 SAIL
06 SASH
07 SACK
08 SAFE
09 SOAP
SOAP
500 WELL
600 WITCH
700 WICK
800 WAVE
900 WEB

The number key has many uses for remembering items in and out of sequence. For example, we could learn this list of musicals:

NUMBER	KEY	ITEM	ASSOCIATION
31	Mat	My Fair Lady	____________
32	Moon	Student Prince	____________
33	Mummy	Oklahoma	____________
34	Mare	West Side Story	____________
35	Mail	King and I	____________
36	Match	Sound of Music	____________
37	Mike	South Pacific	____________
38	Muff	Music Man	____________
39	Mop	Camelot	____________
40	Race	Mary Poppins	____________

Or perhaps you wish to remember a list of books:

NUMBER	KEY	ITEM	ASSOCIATION
21	Net	Treasure Island	____________
22	Nun	Gone With the Wind	____________
23	Name	Ivanhoe	____________
24	Nero	Pinocchio	____________
25	Nail	Baby Care	____________
26	Notch	Bible	____________
27	Nook	Alice in Wonderland	____________
28	Knife	The Robe	____________
29	Nap	Mother Goose Rhymes	____________
30	Mess	Tom Sawyer	____________

7

> "Neither age nor intelligence is a key factor in the ability to remember."
>
> Don A. McKinley

HOW TO LINK AND KEY NUMBERS

In the chapter on the NUMBER ALPHABET, you learned how to remember the number of bones in the body, historical dates, and geographical data. You are now ready to learn how to rapidly memorize longer numbers that have more practical value to you in your business and social life. Such numbers as telephone numbers, addresses, price lists, catalog numbers, math tables, and many others.

In your number key words you have the 1-100 series, 0 to 09 series, and the 100-1000 series. The first two series give you every one and two-digit number. The 100-1000 series is an easy way to expand the Number Key words from 100 to 1,100.

In order to link number key words you simply associate, or link together the key words representing the numbers you want to remember. This is exactly what you do when memorizing a deck of cards in sequence. When you memorize a deck of cards in sequence you are actually memorizing one number of <u>104 digits</u>.

EXAMPLES ON HOW TO LINK NUMBER KEY WORDS

1. <u>Number 386</u>. This number could quickly be remembered by taking the key word for 300, which is WHAM, and the key word for 86, which is FISH, and linking them together. Linked together they become WHAM-FISH.

2. Number 3,672. Translated and linked, it becomes MATCH-CANE.

3. Number 51,907. This is WELL-TAPE-SACK, or if you prefer, it could be HALL-TAPE-SOCK, or even LIGHT-WEB-SOCK.

4. Number 3130886705. This is: MAT-MESS-FIFE-JACK-SEAL.

PRACTICE EXERCISES

	NUMBER	LINKED KEY WORDS
A.	141	______________________
B.	2463	______________________
C.	95612204	______________________
D.	400697	______________________
E.	77828900611239645136	______________________

The above method of remembering numbers is very fast, and primarily designed for remembering numbers only for a short time. For more permanent memory, the invention of words, as we did in the NUMBER ALPHABET chapter, is generally more satisfactory.

KEYING NUMBERS

Keying numbers, which is tying a number to an object, item, or Letter Key, is extremely practical for many purposes. Catalog numbers, stock numbers, telephone numbers, are prime examples. For instance, when you remember a person's telephone number you are actually using him for a key and applying the number to him.

To illustrate the principle of keying numbers, we will learn the Magic Memory Square. This is a very easy thing to do, yet to an audience, a seemingly impossible memory demonstration. After you have learned the Magic Memory Square, you can use it to amaze your friends, and practice this important memory method at the same time.

First you must memorize the chart illustrated on page 44.

You can easily see that the first square is HAY-MESS, square is BEE-TON, etc.

To demonstrate this to an audience, have the chart divided into 25 squares, on a chalkboard or sheet of paper, and the squares lettered from "A" to "Y." However, the numbers are not inserted at this time.

As you look at the chart, the number 100 is the total of the five numbers in the following order:

1. All five columns down.
2. All five rows across.
3. The diagonal from "A" to "Y."
4. The diagonal from "U" to "E."
5. The cross in the middle, such as "L," "H," "N," "R," and "M."
6. The corners plus the middle, such as "A," "E," "U," "Y," and "M."

In order to make the demonstration more effective, we must give the audience a choice of any number, from 60 to 140, that is divisible by 5. It is a good idea to have this number written on the edge of the chart so the audience will not forget it. Make sure that the person who gives this number will verify that you did not prearrange it.

If the number given is 100, you will already have it memorized; therefore your demonstration will be very easy. But, if the number given is below 100, you must use a simple mathematical formula. Subtract the number given from 100, and divide by 5. The answer must be subtracted from each chart number that you have memorized on the chart. If the number given is above 100, you subtract 100 from the number, and divide by five. This answer must then be added to every chart number.

In performing the demonstration, it is a good idea to have someone from the audience writing on the board or sheet. Now, have people in the audience name a letter, and then tell the person at the board which number to write in the corresponding square.

Of course, the audience will call the letters out of sequence, but that will make no difference to you. It is a good idea for you not to look at the board while the letters are being called. To the audience this makes the demonstration appear more difficult.

When the board is completely filled with numbers, you can explain to the audience the six directions which add up to the chosen number.

EXAMPLE: The number given is 70.

100
-70
30 divided by 5 equals 6

Square S is called. Your thinking would be:

S – ESKIMO -TAIL – 15 minus 6 equals 9

You would then say, Square "S" is 9.

The rest of the chart is completed in this same manner.

MAGIC MEMORY SQUARE

A 30	B 12	C 14	D 21	E 23
F 18	G 25	H 32	I 9	J 16
K 11	L 13	M 20	N 27	O 29
P 24	Q 31	R 8	S 15	T 22
U 17	V 19	W 26	X 28	Y 10

8

> "It is much easier to blame your memory than your judgment."
>
> A Wise Man

STOP FORGETFULNESS AND ABSENT MINDEDNESS

A policeman flagged down a San Diego City bus one afternoon and there on an empty seat – right where the owner had left it – was a straw handbag containing $1,714.56 in cash, and $160 in checks. The lady had left the bus, forgetting her handbag.

A Florida man put $890 in a water heater. Later, returning home, he lit the heater to take a bath – then he remembered where he had hidden the money.

A Milwaukee attorney reported to the police that his car had been stolen. The next day he sheepishly informed the police that he had put his car in a garage for repairs and forgotten it.

In Naples, Italy, a professor spent 10 years designing and building a midget submarine. On the first test dive, the submarine sank – because he forgot to close the conning tower hatch.

Every year millions of items are "lost", simply because their owners have forgotten where they left them.

How many times have you forgotten your wallet, or failed to mail an important letter? Have you ever forgotten where you parked your car?

If you have difficulty in remembering common, everyday errands or chores, or are a little absent minded, cheer up. The tips you are about to learn in this chapter will enable you to overcome this problem.

Absent mindedness is usually caused by laziness. We usually forget simply because we haven't made a conscious effort to remember. Haste and tension are also causes of forgetfulness, so slow down a bit and relax. Forgetting is not due to a lack of intelligence or even old age, but rather a lack of being aware when performing an action, or unconsciously acting. We see with our eyes; we observe with our minds. If your mind is "absent" when performing an action, you are not aware.

In order to stop forgetfulness, and absent mindedness, you must make up your mind that you are definitely going to stop being mentally lazy, and that you are going to make a strong effort to remember, by registering something in your mind at the moment it occurs.

The following rules will help you overcome mental laziness and make remembering fun, easy, and time saving. Each rule is explained by an example which you may adapt to your own problem.

RULE 1 – IMAGINATIVE ASSOCIATION

PROBLEM: You have to stop by the cleaners on the way home from work and pick up your suit. In the past you have forgotten to stop at the cleaners, and had to make a special trip back for your clothes.

SOLUTION: When you get in your car, imagine your suit wildly flapping on the hood of your car. Be sure to make a strong mental picture of the suit on your car. If the picture is imagined strongly enough, you will constantly see your suit before you on the hood as you drive home. Don't worry about impairing your vision as you drive – this is merely a mental picture that you are imagining!

By seeing this picture constantly before you, it will be almost impossible for you to forget to stop at the cleaners. As soon as you have picked up the suit, the mental picture will disappear.

PROBLEM: You are just about to doze off to sleep on Tuesday night, when you are reminded that you failed to put the garbage out for collection. You must remember to do this the very first thing Wednesday morning.

SOLUTION: Think that when you place your foot into your slippers in the morning they will be full of wet garbage. If your picture is strong enough, you will really get a jolt in the morning when you put on your slippers, and you will remember to take care of the garbage.

You have learned how to make associations, so it should be easy to apply the rule of association. Just see your object to be remembered in relation to something you have to work with, or use, in your normal daily activity. By so doing, you will be constantly reminded of the item or errand you wish to remember.

RULE 2 --HOOKS ON DOORS

This rule is similar to rule one, except that you use a definite door knob which you are bound to touch after each day. The object to be remembered is placed on the door knob (mentally, of course), and before you can use the door knob again, you have to mentally remove the object which is attached to it. By so doing, you are reminded of your errand or chore.

The HOOKS ON DOORS rule has an added advantage in that you can remember several items by simply linking errands to be done.

For example, on your way home you must stop at the bakery and pick up doughnuts for breakfast. Therefore, you place an enormous donut on your special door knob. Later in the day your daughter calls and asks you to pick her up at dancing class. So you mentally have your daughter dancing on the donut which is on the door knob. Other items can be linked as they occur during the day. When you leave the office, hesitate at the door, and you will see the donut and your dancing daughter, and any other items which you have added to the chain. You can now happily go on your way remembering the things you must do by simply mentally reviewing your mental chain based on your HOOKS ON DOORS rule.

RULE 3 – MENTAL CHECK LIST

When you leave your home, office, friend's house, or wherever you happen to be, pause and think for a moment. Ask yourself: Where am I going now? Have I left anything? What am I to get here to take with me? Do I have my watch, pen, car keys, lunch, brief case? Is everything taken care of? Are the lights off, gas out, windows closed?

Perhaps you are troubled by running out of gas frequently, because you forgot to check your fuel gauge before you started out. If so, develop the habit to mentally check all the gauges in your car when you first start the engine each time.

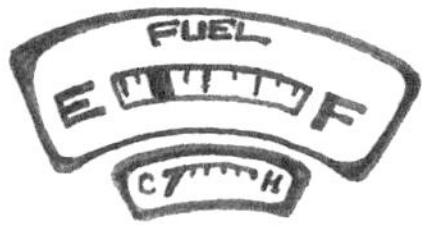

Of course the above questions are just an example of the type you may ask yourself. If this is where you are likely to have the most trouble with forgetfulness, then make yourself a simple checklist and use it often. You will be amazed at how your forgetting problem will be quickly solved.

RULE 4 – A PLACE FOR EVERYTHING AND EVERYTHING IN ITS PLACE

This rule is old, but it is "as good as gold." Don't get in a habit of going into your house and throwing your car keys in the first place you see, and your wallet in another spot, and your pen in still another place. This not only makes for sloppy housekeeping, but it also plays havoc with remembering. The solution, of course, is to have a drawer or a certain place to put everything. Hence, "a place for everything and everything in its place."

RULE 5 – THE DISPLACEMENT OF HABIT PATTERNS

Remember the old story about tying a string around your finger to remember an errand? We certainly don't expect you to tie a string around your finger, but in reality, this principle is an outstanding method in aiding your memory. You can use it many times every day. For example, you must remember to tell your boss about the change in a price list. To use the DISPLACEMENT OF HABIT PATTERNS rule, you can place your ring on a different finger, or turn your chair around when you go out of the office, or leave the waste basket in the middle of the floor. When you see your boss, the displaced object will remind you to tell him about the price list.

The Displacement of Habit Patterns rule is just that simple. Whenever you wish to be reminded of something, just move an object out of its customary place and it will serve as a reminder. Try it, and you will find that this rule really works.

Put a plant on the chair– or the receiver upside down to remind yourself – "order flowers for mother!"

9

> "So if you want people to like you, remember that a man's name is to him the sweetest and most important sound in any language."
>
> Dale Carnegie

NAMES AND FACES

You must MOTIVATE YOURSELF to remembering people. You must BE INTERESTED in people. You must HAVE CONFIDENCE that you can REMEMBER PEOPLE. Although the following is entitled "The Rules for Remembering Names", it is actually the rules for remembering the WHOLE PERSON. Now proceed to read and understand these rules in detail:

I REMEMBER PEOPLE'S NAMES BECAUSE I ALWAYS:

1. Get the name correctly.
2. Repeat the name immediately.
3. Associate the name with something or somebody that I know.
4. Observe the person's outstanding features.
5. Review often enough to keep this person in my mind.

THE RULES FOR REMEMBERING PEOPLES NAMES

These are the rules that you must follow in order to address a person by his correct name.

RULE 1 – GET NAME CORRECTLY

This elementary but necessary rule is very easy to follow. When you meet a person, make sure that you understand the correct pronunciation of his name. You might have to ask this person to repeat his name 2 or 3 or sometimes 4 times, but you must do it. If you fail to do this you certainly will be embarrassed later.

RULE 2 – REPEAT THE NAME IMMEDIATELY

You can't do this rule if you didn't do Rule 1. By repeating the name you are putting it in your name vocabulary correctly pronounced. This not only helps your memory but it is an excellent way to improve your human relations. It is an established fact that almost everyone likes to hear the sound of his name more than any other sound. Research has shown that when people meet, the first impression is a very lasting one. Therefore, there is no better way to help make people like you, than to make a good first impression by calling a person by what he likes to hear, his name. Of course, you add his name to a phrase such as "it is nice to meet you," or "it's indeed a pleasure," or "I'm glad to know you." You must do these two rules before you start into a conversation.

RULE 3 – ASSOCIATE THE NAME WITH SOMETHING OR SOMEBODY THAT YOU KNOW

Associate means to put information into your mind by placing it with information that you already have. This is the basis of almost all learning, but most people associate without knowing they're doing it, and without any training, and therefore they do a poor job of it. Let's consider the three categories of names.

In the FIRST CATEGORY are the names that actually mean SOMETHING. People are named for colors, places, directions, occupations, hobbies, animals, plants, the weather, moods, and all kinds of objects. When you meet a person with a name that means something to you, then you think of this person either doing or having to do with, whatever the name means.

Meet Mr. BARR.

You have just met him so you know nothing about him. Therefore you create your mental association purely on your first impression of him. You might imagine him as a lawyer pleading his case before the BAR.

Possibly you would think of him as owning a BAR – BARR's BAR. Maybe you think he is rather stiff and formal, like a CROWBAR. It makes no difference that he is none of these things. Whatever association you use, you exaggerate it, animate it, so that it stands out vividly in your MENTAL PICTURE.

An insurance agent (but a real name loser) met an excellent prospect and sold him on a large policy. The conversation ended like this. "Come by my house tonight. Bring the application and I'll have a check ready for you."

"Your street number?" the agent asked nervously.

"It's by my name in the phone book, so long."

Here is Miss VIOLETTE. VIOLET can be a flower, a scent, or a color. Perhaps the easiest one to fix in your mind is the flower. Don't just see the flower with a face in it, but see this flower dancing through a field of grass or a big pasture. Put a lot of fast action and imagination into the picture in your mind.

MISS VIOLETTE

MR. CRANE

This is Mr. CRANE. You might imagine him as a mechanical CRANE, or even as operating the CRANE. Also, it is possible to envision him as a bird, CRANE, walking around. Whatever association you use, make it live in your mind. The more ridiculous and animated the association, the easier it will be to record it in your mind for future RECALL.

Next I will introduce Miss GARDNER. This name suggests one thing so strongly that you need not look for any other association. Picture her in the GARDEN working with her plants and flowers. Does she look like a real GARDNER? If you met her for the first time in her GARDEN clothes, would you recognize her at a party?

MISS GARDNER

MR. MOORE

This gentleman may be a little harder to visualize as his name is not that of an object, animal, or an occupation. You could, however, think of MORE. There may be MORE of him, tall or broad, he may be MORE talkative or maybe MORE handsome. Look him over and see what he has MORE of, then build your association on that item.

EXERCISE #1

This exercise will help you determine how familiar you are with FIRST CATEGORY names. Some names might come under more than one area, but only use any name once. Write on the blanks a surname that can be described by each of the following.

COLORS ______________________________
PLACES ______________________________
DIRECTIONS ______________________________
OCCUPATIONS ______________________________
HOBBIES ______________________________
ANIMALS ______________________________
PLANTS ______________________________
WEATHER ______________________________
MOODS ______________________________
OBJECTS ______________________________
OTHER WORDS ______________________________

Several possible answers to these are listed at the end of the chapter.

Included in the SECOND CATEGORY are names that already belong to SOMEONE else. You do not know what the name means but you already know somebody else with the same name. This somebody can be a personal friend, a movie or T.V. star, an athletic hero, a statesman or even the character in a book or play. You will find that the majority of names you know will come in this category, and that you know a lot of names that come in both the first and second categories. When you meet a person with a name that means somebody to you, then you associate this new person with the one that you already know.

There are several BOONES you might use for your association, even though their names are spelled with an E on the end. There is DANIEL BOONE, the famous pioneer, and PAT BOONE the singer. Or you might think of RICHARD BOONE of "Have Gun, will Travel" fame.

MISS BOON

MR. DAVIS

Mr. Davis could be fixed in your mind by thinking of Bette Davis, the fine movie star. You might see him in a gray uniform as JEFFERSON DAVIS, Confederate States President. Possibly a good association would be SAMMY DAVIS, Jr. the entertainer.

Of course, there is Sir MALCOMB CAMPBELL and his son DONALD, who are world title holders for auto speed runs. Also there is the Scottish poet THOMAS CAMPBELL. Perhaps the most easily remembered name is CAMPBELL'S SOUPS. Try seeing Mrs. Campbell making some alphabet SOUP.

MRS. CAMPBELL

MR. WILSON

A good association for Mr. WILSON would be our late President WOODROW WILSON. Or, how about the buxom entertainer MARIE WILSON. You could picture him jumping rope with a string of wieners from WILSON MEAT COMPANY.

Perhaps everyone would think at once of our late President as the mental association here. Also any of the other KENNEDYS could be used. Perhaps you would recall the character movie actor EDGAR KENNEDY.

MR. KENNEDY

EXERCISE #2

This exercise is an opportunity for you to check your vocabulary of SECOND CATEGORY names. Write the first name of three people with these popular surnames. At least two of these should be of national reputation, while the third can be a personal friend. Brand names of products can also be used.

JONES ____

ALLEN ____

ROGERS ____

ADAMS ____

JAMES ____

BENNETT ____

LINCOLN ____

JOHNSON ____

SMITH ____

WILLIAMS ____

Several possible answers to these are listed at the end of the chapter.

The THIRD CATEGORY are names that do not qualify for either of the other two. You will consider these as LEARNING PROBLEMS and these problems will be fun to solve. When you first hear a learning problem name it will be meaningless, but you will make it mean something by adding, changing, or omitting a letter or some other thinking process that means something to you. Generally you need to say the name, to yourself, very slowly, a few times to get the correct pronunciation. When you do this you will find that the name might mean two or three things, or it might sound like a phrase. It will take you longer to get these names but they will be easy to remember. Once you learn a learning problem name it goes into your second category. As you meet more and more people you increase your vocabulary of names, thus the more people you know the easier it is to remember more.

Although the spelling is different, you could picture a FERRIS WHEEL for mental association. Possibly you remember the little rhyme, "Mirror, Mirror on the wall, who is FARRIS of them all?"

MISS FARRIS

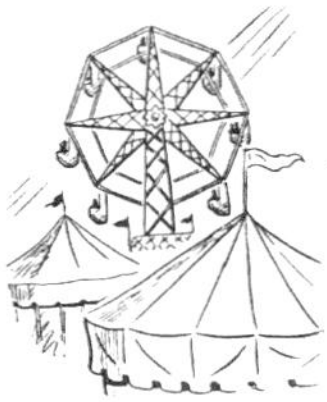

MR. HALGREN

Mr. Halgren might be identified with WALGREEN DRUG STORES. HAL GREEN or GREEN HALL is also a possibility. Use plenty of imagination in slightly distorting this type of name to fit in with something by which you may be able to remember it.

MR. CLABBY

Maybe MR. CLABBY is quite FLABBY. He could be CRABBY, or maybe he is CLUBBY (that is CHUMMY). There are several variations, one of which should strike your fancy as an association that will work for you.

MR. CIMMERRUSTI

MR. CIMMARRUSTI is a good test for your imagination. It could be identified by SIMMER RUSTY or SOME ARE RUSTY, or even by SEE MY RUSTY Of course it must be something with which you can associate the REAL NAME for RECORDING in your mind.

MR. WOJCIECHEWSKI

MR. WOJCIECHEWSKI

This last one looks extremely difficult, but actually it is the easiest of them all. Yes, MR. WOJCIECHEWSKI is a very simple pronounced WATCH A COW SKI. This readily produces a mental picture not easily forgotten.

EXERCISE #3

Let's test your imagination on THIRD CATEGORY names. These names are meaningless as spelled or pronounced, but you make them mean something by CHANGING, PREFIXING, SUFFIXING,

INSERTING OR OMITTING a letter. Often the chosen word will rhyme with the name. Don't be concerned with the spelling, pronunciation is the important thing. Get two words to associate with each name.

SARDEN ______________________________

RUDD ______________________________

LUND ______________________________

HEATH ______________________________

ASCHER
(Ash-er) ______________________________

VOYLES
(Voi-els) ______________________________

SEAVER
(See-ver) ______________________________

MEISEL
(My-sel) ______________________________

Do Exercise #4 before checking your answers.

EXERCISE #4

Make a phrase or sentence that you can turn into a Mental Picture to associate these names. Don't be afraid to let your imagination run wild.

TOWNSEND ______________________________

HARTLE ______________________________

DUNHAM ______________________________

McINDOO ______________________________

PUNTENNEY ______________________________

HOMRIGHAUS ______________________________

SANDSTEDT ______________________________

Check your answers with those at the end of the chapter.

An absent-minded professor entered his home one evening and found a woman lounging on the couch. "Madam," he exclaimed, "What are you doing here?" Said the lady, quite surprised, "I like this couch. I like this house. I like this neighborhood. I like you, and besides, I AM your wife."

LET'S SEE HOW YOU ARE DOING —

Here are the 15 people you have already met, posed differently. Recall and write the correct name under each picture. Then turn back and check yourself. Do not proceed until you know all of them.

RULE 4 --OBSERVE THIS PERSON'S OUTSTANDING FEATURES

Very few people have trouble in remembering faces, it is the names they forget. This is true, because when you talk to a person, you naturally observe their looks, and this is recorded in your mind, then when this person approaches you on another occasion, you recognize who it is and also recall where you met before. If their name was correctly associated with their looks you will be able to call them by name instantly. If you do need help on this rule, then you make an extra effort of thinking of one or more of these features. Their height, weight, posture, hair color, complexion, or whether head, eyes, ears, nose or mouth are large, small or in anyway unusual. Do not think of things that can be changed such as clothes, glasses or makeup, and do not think of this person only in this environment. Do not be misled into thinking that the name and face can be associated as being one and the same. When you put the name and the person's looks in your mind at the same time, properly, you will remember the name when you see the person.

Many readers will not need to be concerned with this rule because it is a fact that the vast majority of people DO REMEMBER FACES. This does not mean that they do not follow the rule, it means that they do it without any effort. Some people definitely need help on this rule. A thorough understanding of the CHART OF PEOPLE'S OUTSTANDING FEATURES on the following page will eliminate this problem.

There are close to three billion people in the world and no two of them look exactly alike. Even identical twins can be distinguished from each other when you observe them closely. In looking for features, OUTSTANDING means what it says, look for those that are really distinctive and not what seems average.

You might try to guess this person's AGE, HEIGHT, and WEIGHT, and, if you do find the correct answer to these, then you probably will become a good guesser. Certain nationalities have somewhat distinctive features and if the person's name is indicative of that nationality it will help you. The mixture of nationalities in the United

States makes it so that just as often as not the person WILL NOT look like the nationality his name would indicate. This is helpful as an opposite association.

Study the CHART occasionally until recognition of one, two, or three of the person's outstanding features comes to your mind without much mental effort when you are meeting them. When you do this properly you will find that you will be able to recognize them, and call them by name instantly.

Facial Features and Body Types

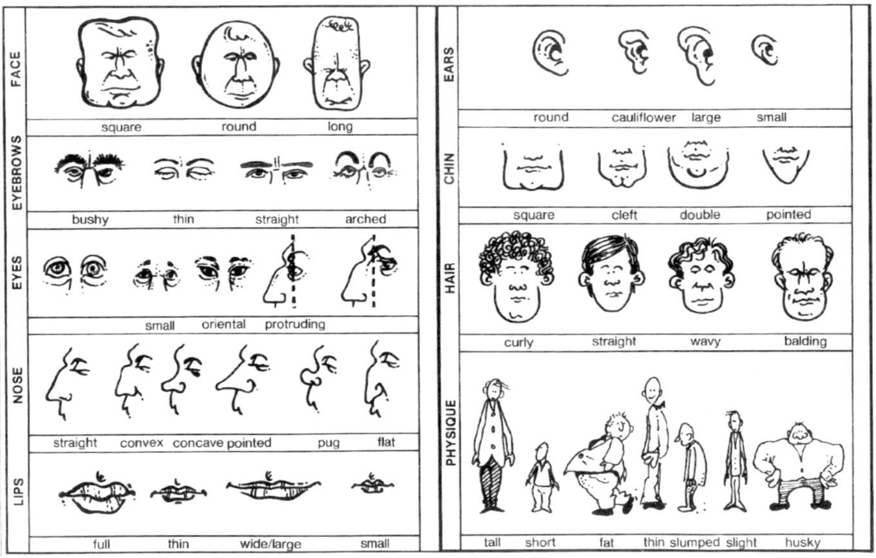

RULE 5 --REVIEW OFTEN ENOUGH TO KEEP THIS PERSON IN YOUR MIND

How often is often enough? This is quite variable. It depends upon your motive, your interest, your natural memory, but most important on how well you associated the name.

When you are interested, or have a motive to remember people, you will review by thinking about them, talking about them, and talking to them. Time is an enemy of your memory, but when you follow the rules you overcome this enemy. Keeping rosters, keeping a file of people's business cards, having a record of customers or clients in a notebook or file cabinet, is good business. Take advantage of these by spending some time reading through them. When you think of a person, run through your mind for much of your knowledge about him. Think of him as a person and why you like him as a friend. Think of his name, features, business, hobbies, interests, and anything else you know about him. Think of the sound of his voice, and you will be able to spot him when he telephones you. The more you repeat this information the more firmly it is established in your mind and the easier and quicker it is to recall when the occasion arises.

These five rules are natural and everyone follows them whenever they remember a name, even though they are not aware of doing so. Whenever someone forgets a name then one or more of the rules was not followed. You will develop the habit of following the rules properly, and you will remember the names of all the people you wish to remember.

When you have applied all of the RULES, you will have OVER-LEARNED your knowledge of your friends. Then SUCCESS THROUGH REMEMBERING PEOPLE is yours.

GIVEN NAMES ARE EASY —

Given names (the first and middle name) are generally easier to remember than surnames. This is true, even though most of them do not have a meaning. One reason is that about 100 female and 100 male given names are shared by the vast majority of people. Therefore, you keep meeting

more and more people with the same given name, as compared with thousands of surnames. Another reason is that you get more practice with the given names because in your social life entirely and also quite often in your business life you address people by their given name.
This is what you must do to assure yourself of remembering the entire given and surname

WHEN YOU DO RULE #1,
GET THE GIVEN AND THE SURNAME

WHEN YOU DO RULE #2, CALL THE PERSON BY THE FIRST AND/OR/LAST NAME OR BOTH, WHATEVER IS MOST APPROPRIATE FOR THE OCCASION, BUT THINK OF BOTH GOING TOGETHER.

DO RULE #3 AND RULE #4 AT THE SAME TIME. ALTHOUGH MOST OF THE ASSOCIATING WILL BE WITH THE SURNAME, BE SURE TO THINK OF BOTH NAMES TOGETHER, AND SAY THEM TO YOURSELF A FEW TIMES TOGETHER WHILE YOU OBSERVE THIS PERSON'S FEATURES.

When you follow these rules, the given name, the surname and the person's looks go into your mind together, the correct way. When you later think of this person, or recognize him, then both names will come to your mind and you will address him by name, quickly and correctly.

Look for connections that make the given and surname go together and it will be a big help to you. Here are some examples of rhyming names. JUNE MOON, WALTER WALKER, PAUL PAULSON, and BETTY BERRY. In these examples the combination of names makes a meaning. VIRGINIA REAL (the dance), TOM COLLINS (the drink), BEVERLY HILL (the city), SHERRY WINE (the drink), PAUL BARRER (the duty), GUY WEIR (the support), LES MOORE (is it less –or more?), HAZEL BROWN (yes, hazel nuts are brown), TED(dy) BEAR (isn't he cute), GALE RAINS (usually gales bring the rain), JUNE MAY (she has her months backwards), OTTO KARR (of course it's auto-car), and HARRY HYDE (like a monkey).

If the initials are the same, or in alphabetical order, then make a mental note of it, such as BOB BLISS, SAM SMITH, MARY MILLER, or AL BRIGGS and ROGER S. TAYLOR. Sometimes the initials spell a word as DONALD O. GODDARD (DOG), JUDY ANN MASON (JAM), and RUDY A. WARD (RAW). Occasionally the initials match some famous initials, such as DONALD A ROBINS (Daughters of the American Revolution), FRANK D. READ (F. D. ROOSEVELT), and VICTOR F. WHITE (Veterans of Foreign Wars).

Just like FIRST CATEGORY surnames, there are some given names that mean something. When they do, take advantage of the fact. Most given names will be easy SECOND CATEGORY names. Given names in the THIRD CATEGORY are learning problems to be solved just as you would with surnames, and they are easy *too.*

Exercises #5 and #6 are on the following pages. Follow the directions and be sure to do every name. Of course, some will come easy and others will make you do some memory searching. There are no answers for these exercises because most of your answers will be friends of yours.

EXERCISE #5

A partial list of the most popular female given names is listed below. Beside each name, write the surname of someone you know. This person may be a friend, relative, celebrity or public figure. There might be a difference in spelling, or a nickname from these.

ALICE	____________	EDITH	____________
ANNE	____________	EDNA	____________
BARBARA	____________	GLADYS	____________
BETTY	____________	GLORIA	____________
BEVERLY	____________	GRACE	____________
CAROLINE	____________	HAZEL	____________
DEBBIE	____________	HELEN	____________
DOROTHY	____________	IRENE	____________
JANE	____________	MARILYN	____________
JANET	____________	MARTHA	____________
JEAN	____________	NANCY	____________
JUDY	____________	NORMA	____________
KAREN	____________	PAMELA	____________
KATHERINE	____________	PATRICIA	____________
LAURA	____________	PHYLLIS	____________
LINDA	____________	RUTH	____________
LOUISE	____________	SHIRLEY	____________
LUCY	____________	SUSAN	____________
MARY	____________	VIRGINIA	____________

Overheard in a marriage counselor's office, "Well, to start with, what's-her-name here, my wife, claims that I am not interested in her."

EXERCISE #6

A partial list of the most popular male given names is listed below. Beside each name, write the surname of someone you know. This person may be a friend, relative, celebrity or public figure. There might be a difference in spelling, or a nickname from these.

ALFRED	____________	DONALD	____________
ANTHONY	____________	EDWARD	____________
ARTHUR	____________	EUGENE	____________
CARL	____________	FRANK	____________
CHARLES	____________	FREDERICK	____________
DANIEL	____________	GEORGE	____________
DAVID	____________	HAROLD	____________
HARRY	____________	PETER	____________
HOWARD	____________	RALPH	____________
JAMES	____________	RAYMOND	____________
JOHN	____________	RICHARD	____________
JOSEPH	____________	ROBERT	____________
LAWRENCE	____________	ROGER	____________
LLOYD	____________	RUSSELL	____________
LOUIS	____________	STEPHEN	____________
MICHAEL	____________	THOMAS	____________
NORMAN	____________	TIMOTHY	____________
PATRICK	____________	WALTER	____________
PAUL	____________	WILLIAM	____________

EXAMPLES OF USING THE RULE PROPERLY

- As you do Rule 1, ask for a business card or ask how their name is spelled.
- Do Rule 2 immediately. Say something like "It's a pleasure to meet you Mr. Shively. I am interested in surnames, what nationality is Shively? I have never known the name before." He tells you that it is German, but he doesn't know the meaning.

- Do Rule 3 by thinking along these lines, as fast as you can. It's spelled DON SHIVELY. Sounds like SHove chIVes for LEE. Say to yourself a few time DON SHY Va Lee until you are sure you have it right. This will take you from 1 to 10 or maybe 20 to 30 seconds, depending on how fast your mind can comprehend this new name.
- Do Rule 4 by looking right into his eyes and at his face (you don't have to stare) while you are doing the first 3 rules. You might think something like WHITE HAIR -ROUND FACE - HIGH FOREHEAD.
- Do Rule 5 by saying his name to him a few times during your conversation and upon saying goodbye. Put his card (or a card that you write his name and a few facts about him) in your Name Card File. Tell your friends about your meeting him. Think about him occasionally. Go through your Name Card File periodically.

Be INTERESTED in him as a new friend of yours.

Have CONFIDENCE IN YOUR ABILITY to remember him.

WHERE DO NAMES COME FROM?

An understanding of the derivation of names is helpful in remembering them. A short discussion of the meaning of a new acquaintance's name helps you learn more about names, and virtually assures you of remembering this person's name. It also convinces this person that you are interested in him, a most important step in making friends. In researching the derivation of names, the inexact science of onomatology, I have found many contradictions. These are caused by lack of records, the changing of names, a change in the spelling or pronunciation of the name and other reasons. This just adds to the fun of remembering people.

Previous to 1000 to 1500 AD, most people just had one given name, especially in the European countries (this is still true with some cultures). The time when surnames came into use varied considerably from country to country and even from area to area. Generally, as more and more people lived in or near a village or city, it was necessary to add a surname to distinguish different people with the same given name. Usually surnames were derived from one of the following four sources.

OCCUPATIONAL NAMES: This was one of the easiest ways to distinguish a person, because everyone in the area would know who it was. Most of the names listed in the answers to Exercise #1 OCCUPATIONS, are of English origin and also some of the names listed in HOBBIES and probably some from other groups were from this source. Of course, the name Smith (probably the most popular name in the United States) is an occupational name.

LOCALITY AND SPOT NAMES: These derive from the place where a person lives or something prominent about his dwelling or area. In the answers to Exercise #1, not only.PLACES but DIRECTIONS, COLORS, WEATHER, OBJECTS, and some of the OTHER WORDS would come from this source. Examples: the family up on the HILL, the man with the big ROOME, with the GRAY barn, EAST of town, he likes the RAIL or they came from HOLLAND. Many ANIMAL and PLANT names come from tavern signs or shop signs used before there were street numbers.

NICKNAMES: These depict a peculiarity, deformity, style of dress, and may have been either complimentary, humorous or even ridiculous. Like nicknames you pick up from your friends, they often stick. In the answers to Exercise #1 MOODS, and some of the OTHER WORDS are names from this source. Example: he is a MOODY person, they are a DARLING family. Names such as KING, QUEEN, EARL, and PRINCE were probably nicknames of popular persons. Most of these examples are English names but the same principle applies to most languages. If you don't speak that language it doesn't help you, but by asking a person the meaning of his name helps you to build your name vocabulary.

PATRONYMICS (PAT ro NIM ics): This means adding a prefix or suffix to the father's given name, that means_SON OF. The biblical name BARJONA means Son of Jona. In English this was done by adding SON such as JOHNSON, or just adding as S such as ROBERTS. Actually JONES is a short way to spell Son of John. This was done in many languages. Here are a few examples. Prefixes 0' in Irish such as O'BRIEN, Mc or Mac in Scottish and sometimes Irish such as McDONALD. FITZ is from Latin and used in French and some English names such as FITZSIMMONS which means Son of Simon. In Welch this was done by prefixing Ap which means Son, but also shortening the name, such as PRICE (ap-Ryhs) and PUGH (ap-Hugh). Some common suffixes are SEN or SON in the Scandinavian countries, SOHN or MANN in German, VICH or WITZ in some Balkan countries and OV or SKI in Russian and Polish.

Here is an example of how these names can get all mixed up. In German, berg means a hill or mountain, so a man takes the surname BERG because of his dwelling being up in the mountains. His son adds the suffix mann so his surname becomes BERGMANN. Now another BERGMANN moves near the village but on a lower slope of the mountain. The villagers, to distinguish between the two, call the upper mane OBERBERGMANN and the lower one NEDERBERGMANN. When someone of such a long surname moves to another area or country, he drops some of the letters and/or the suffix or prefix. This was done quite often with names and are hard for people in another country to pronounce, such as the mixture of nationalities and races moving to the United States. Many surnames have been altered so much through the years that they have lost their original meaning. The only names that actually originated here are those of the Indians. Many tribes used names such as RUNNING FOX, MIGHTY EAGLE, DEAD HORSE, or GREEN BRANCH for given names and parts of some of these probably are now surnames of their descendants.

The custom in Scandinavian countries is for the suffix SON or SEN to be added to the father's given name instead of the surname.

Therefore, a man names JOHN OLSON, would be the son of OLE PETERSON and the grandson of PETER SVENSON. In Spanish speaking countries there is a custom of adding thc mother's maiden name following the actual surname. For example FRANCISCO DIAZ GARCIA should be addressed Mr. Diaz and not Mr. Garcia. Chinese surnames are always pronounced with one syllable such as WONG, HOM and CHU. Japanese names are easy to pronounce because they usually follow the pattern of consonant-vowel-consonant-vowel etc., such as MORIMOTO or NAKAMURA. Many original Greek names end in POULOS, which means name. In the United States many Greeks have changed their name to make it easier to pronounce because in Greece the names are very long. Quite often a surname that ends in I or 0 is of Italian origin. VAN is Dutch and VON is German. Whenever they are in front of the surname, they were titles bestowed by royalty.

In addressing a person, be careful to use correct pronunciation, which is the way this person wants to be called. For example, the name KNUDSON- some owners prefer to have the K pronounced and others do not. Although from the same family, our two presidents ROOSEVELT preferred different pronunciation of their names. It was Theodore ROOSE-VELT and Franklin D. ROHS-A-VELT. This does not cover all there is to know about names, but these ideas will be of help to you in remembering people.

MEETING GROUPS OF PEOPLE

When you are going to attend a party, dinner, convention, or join a new church, school, club, or any other situation that you might be meeting many people at one time, it is important to do these things.

OBTAIN A ROSTER or list of those who will attend and study it in advance. Learn to pronounce the names before you meet the people. Be on the lookout for the names that you have studied. As you are introduced, it is good business to also know something about these people, which gives you a lead into a conversation.

GET THERE EARLY SO YOU CAN MEET THE PEOPLE INDIVIDUALLY OR IN SMALL GROUPS AS THEY ARRIVE. This allows you more time to properly meet the people.

DON'T BE RUSHED beyond the point of following RULES 1, 2, 3, and 4 of Remembering People's Names. If you do, you will surely be embarrassed later. Let your host or hostess know that you really want to KNOW these people, not just meet them.

REVIEW (Rule 5) at every opportunity, the names of everyone you have just met, by looking around, and testing yourself. This establishes them more firmly in your mind. If you miss some, ask a friend, or refer to your roster. If you don't have a roster then get off to yourself and write their names down on a card.

ALWAYS USE THE PERSON'S NAME in speaking to, or referring to him, and in saying goodbye. The following day, think again of these new friends of yours.

Don't be discouraged if you are not 100% perfect, remember Babe Ruth didn't hit a homerun every time. When you practice properly you will continue to get better and better. As you meet more and more people you increase your vocabulary of names, thus the more people you know the easier it is to remember more.

Always wear your own convention or party name tag upside down. Then if someone asks you your own name, you won't have to get flustered, you can look down and read it.

Take them one at a time and read the information about the person. Look at the picture, and follow the Rules just as if this person were actually talking to you. Continue with the remaining pictures on the sheet until you have all of them, then test yourself. Keep these with you for a week and test yourself at least once a day. Each succeeding week take the next sheet and repeat the process. Continue to test yourself on the ones you have already learned.

When you reach the last sheet you should be able to meet the people quite rapidly. The more you practice, the easier it will be for you to apply the Rules to everyone you meet.

GOOD LUCK TO YOU

Now you have the knowledge, the interest, and the self-confidence to remember the names of all the people you wish to remember. Putting these to work brings you satisfaction and SUCCESS THROUGH REMEMBERING PEOPLE.

ANSWERS

ANSWERS TO EXERCISE #1

COLORS:	White. Black. Red. Green. Blue. Brown. Gray. Violet. Orange. Coral.
PLACES:	Hill. Hall. Park. Fields. Wall. Roof. Roome. House. Beach. Pound. Woods. Works. Grave. Ward. Holmes.
DIRECTIONS:	North. East. West. South. UpP. Downs. Forward. Back. Lowe.
OCCUPATIONS:	Skinner. Foreman. Dyer. Gardner. Baker. Barker. Dorman. Banks. Barber. Trainor. Binder. Miller, Glazier. Parsons. Elder. Outlaw. Crooks. Hood. Begg.
HOBBIES:	Hunt. Fisher. Stamper. Coyne. Clymer. Goff. Horn. Flyer. Camps. Trapp. Box. Player. Card. Ryder. Sailor. Learner.
ANIMALS:	Pike. Trout. Herring. Salmon. Bear. Fox. Lyon. Wolf. Robins. Martin. Wrenn. Jay. Partridge. Pidgeon. Hogg. Lamb. Collie. Kidd. Catt.
PLANTS:	Bush. Ivy. Rose. Berry. Apple. Cherry. Flowers. Laurel. Mellon. Cotton. Forrest. Root. Branch. Thorn. Astor.
WEATHER:	Gale. Storm. Wind. Cloud. Rain; Snow. Rainwater. Brooks. Lake. Rivers. Oceans. Sea.
MOODS:	Gaye. Moody. Frank. Earnest. Hope. Madden. Wise. Love. Hayter.
OBJECTS:	Post. Cole. Boote. Star. Dolley. Stone. Rhodes. Gates. Moon. Rope. Bell. Light. Glass. Jewel. Hoe. Carr. Coates. Nichols. Iron. Gold. Silver.
OTHER WORDS:	Sharp. Short. Day. Long. Noyes. Biggs. Pierce. Little. Seymore. Heard. Stout. Frey. Hitt. Young. Olds. Summers. Winter. Spring. Fall. King. Queen. Savage. Armstrong. Fleet. Flint. Irish. Britten. German. Berlin. Paris. Holland. Heater. Burns. Angel. Poor. Rich. Freeman. Miles. Swift. Read. Darling. and thousands more.

ANSWERS TO EXERCISE #2

JONES:	John Paul-Father of U.S. Navy. Casey of Railroad Fame. Bobby-Golfer. Shirley-Singer-Actress. Parnelli-Race car driver.
ALLEN:	Steve-Multi-Talented T.V. Star. Ethan-Revolutionary War Hero.

ROGERS:	Roy-Famous Cowboy, Ginger-Dancer-Actress, Will-Cowboy and Humorist, Buck of the 25th Century, Richard-Composer.
ADAMS:	Samuel-Father of the American Revolution, John-2nd President, John Quincey-6th President, Family-T.V. Program.
JAMES:	Harry-Band Leader, Jesse and Frank-Train and Bank Robbers, William-Philosopher and Psychologist, King James 1 and King James 2-Kings of England 15th Century, The Name of Two of Christ's Disciples.
BENNETT:	Millard-Expert on Salesmanship, Wallace Foster-Senator from Utah, First Name of Bennett Cerf-T.V. Panelist.
LINCOLN:	Abraham-16th President, the Automobile, Victoria-Writer, the Capital of Nebraska.
JOHNSON:	Lyndon B.-36th President, Andrew-17th President, the Wax,
SMITH:	Capt. John-Founder of Virginia, Joseph-Father of the Mormon Church, Deaf (Erasmus)
WILLIAMS:	Andy-Singer and T.V. Star, Roger-Pianist, Tennessee-Writer, Tex-Cowboy Singer. By looking in an Almanac, Encyclopedia, and Who's Who you will find scores more.

ANSWERS TO EXERCISE #3

SARDEN:	Garden, Harden, Pardon, Sardine, Sardinia.
RUDD:	Red, Mud, Dud, Bud, Cud, Rut.
LUND:	Fund, Lund, Lend, Lung.
HEATH:	Sheath, Health, Heat, Wreath.
ASCHER:	Hasher, Rasher, Casher, Lasher, Masher, Ashen.
VOYLES:	Coils, Boils, Foils, Toils.
SEAVER:	Beaver, Weaver, Fever, Sealer, Sever, Seven.
MEISEL:	Myself, Mysell, Miser, Nice.

ANSWERS TO EXERCISE #4

TOWNSEND:	Go to the town's end. Send it to town.
HARTLE:	Heart shaped. Hurdle a heart.
DUNHAM:	Cook the ham done. He is both a dunce and a ham.
McINDOO:	Yes Mac can do it. Mac a Hindoo?
PUNTENNEY:	Not a football, punt a penny. What a tinny pun.
HOMRIGHAUS:	A home and a house with a rig in between. Go home and rig the house.
SANDSTEDT:	On the sand instead of the grass. A sandy step.

10

"Memory is invaluable because it is through this retentive facility of acquired knowledge and experience that we develop judgment and skill.

— Paul J. Meyer

REMEMBERING DAILY AND WEEKLY APPOINTMENTS

In Chapter 8 we talked about remembering errands and appointments by using the linking method. However, if you choose to remember your daily appointments for the week, here is a method you might use.

It is easy to do! Refer to the chart on the following page and you will see we have designated the various hours in the day with peg words. For example, Monday is the first day of the week, so the first digit is 1. The second and third digits represent the hour of the appointment. If we had a 4:00 p.m. appointment on Monday, we would associate the name or activity with the peg word TAR or 14. One representing the day, or in this case, Monday. And the four representing the hour.

You would not be able to visualize the number 14, but you can see the word TAR and then associate it with the appointment. If you had a dentist appointment, picture TAR on your teeth or false teeth stuck in the TAR. If it were a name, then link that name with the TAR.

Tuesday	9:00 a.m.	would be	29	or	NAP
Wednesday	11:00 a.m.	would be	311	or	MATADOR
Thursday	3:00 p.m.	would be	43	or	RAM

Qty.	Item
1	**Slaying Dragons and Moving Mountains: The MAGIC Formula for a Happy, Fulfilled Life...** Wissenbach, Stefan --- Paperback **0992636302** 0992636302 9780992636302
1	**Ideal Weights: Attaining Your Ideal While Achieving Lasting Success—Even Post Menopause!** Azim, Dr. Mahnaz --- Paperback **0578723093** 0578723093 9780578723099
1	**Miracle Memory Methods** Mckinley, Don --- Paperback **0982878206** 0982878206 9780982878200
1	**Legal Writing with Style** LeClercq, Terri --- Paperback **0692578153** 0692578153 9780692578155
1	**Money Rules For Women: The Proven Secret System To Master Your Money And Achieve Financial Certainty** Kleinman, Mr Nobby --- Paperback **098071074X** 098071074X 9780980710748

Return or replace your item
Visit Amazon.com/returns

0/Nxy13cTXN/-10 of 10-//MKE6-CART-CO/next-1dc/0/1105-23:00/1105-18:49 **A1-110**

Friday	12:00 a.m.	would be	512	or	ALLADIN
Saturday	6:00 p.m.	would be	66	or	JUDGE
Sunday	10:00 a.m.	would be	710	or	COATS

Familiarize yourself with these key words and then as you make the appointments key them in with the peg word and review each day when you are driving to work, having lunch, etc. Review each day's appointments the night before, but don't forget to review the other daily appointments for the week.

Now your next questions might be, "What about appointments other than on the hour?" Well, link or associate a quarter, or quarter of an apple for 15 minutes past the hour with the key word representing the hour appointment. Use 1/2 of a cantalope or grapefruit for the half hour. And for three quarters or 15 minutes to the hour, use the visual picture of three quarters of a cake or pie, again linking these with the key words and appointment.

For earlier or later appointments, use the color white for a.m. and black for p.m. For example, 6:00 a.m. Monday would be white tissue. Or 6:00 p.m. Sunday would be black cash.

	MONDAY	TUESDAY	WEDNESDAY	THURDSAY	FRIDAY	SATURDAY	SUNDAY
	1	2	3	4	5	6	7
A.M. 7:00	TACK	NOOK	MIKE	RAKE	LOCK	JACK	COKE
8:00	TAFFY	KNIFE	MUFF	ROOF	LEAF	JAVA	CAFE
9:00	TAPE	NAP	MOP	ROPE	LIP	JEEP	COP
10:00	TOADS	NUTS	MEDICINE	ROADS	LETTUCE	SHEETS	COATS
11:00	DDT	KNITTED	MATADOR	RED HEAD	ALTITUDE	CHEATED	CADET
12:00	TITANIC	INDIAN	MITTEN	WARDEN	ALLADIN	SHOOTEN	KIDNEY
P.M. 1:00	TOT	NET	MAT	RAT	LIGHT	JET	CAT
2:00	TON	NUN	MOON	RAIN	LION	JUNE	CANE
3:00	TIME	NAME	MUMMY	RAM	LIME	JAM	COMB
4:00	TAR	NERO	MARE	ROWER	LAIR	JAR	CAR
5:00	TAIL	NAIL	MAIL	RULE	LILY	JAIL	COIL
6:00	TISSUE	NOTCH	MATCH	ROACH	LASH	JUDGE	CASH

11

> "Remembering is everybody's business, every minute. Luckily we can greatly increase -and easily -the amount and accuracy of what we remember."
>
> — Donald A. and Eleanor C. Laird.

SHORT CUT CALENDAR

If you wanted to know what day of the week falls on any date within the year you could refer to the chart on the next page to see which is the first Sunday of the month for each of the twelve months. If you relegate to memory the twelve digit number, or better yet, form a link or sentence with the number alphabet, you will be able, almost instantly, to locate the day of the week you are looking for.

The first Sunday in January 2017 falls on the first day of the month; first Sunday in February is on the fifth; first Sunday in March is also on the fifth; first Sunday in April is on the second; and so on to December, the first Sunday of which is on the third of the month.

Once you know the first Sunday, you can easily calculate the day of the week for any day within that month. Suppose you wanted to know the day of the week for October 23, 2012. The first Sunday in October falls on the seventh. Add two weeks to that digit and you know that the 21st is Sunday. So by adding two days we find that the 23rd falls on Tuesday.

Let's assume you wanted to know the day of the week on which Christmas day 2018 falls. The first Sunday is on the seventh. So, adding two weeks we find that the 21st is Sunday. We add four days which means that Christmas is on Thursday. We also could add three weeks and work backward by subtracting three days.

If your birthday is July 29, and you want to know on what day in 2015 your birthday falls, we look at the chart. In July 2015, the first Sunday is on the fifth. Adding three weeks to this date makes the 26th a Sunday. So by adding three days to this date, the 29th falls on a Wednesday.

Generally you would only need to remember one year at a time. So it would only be necessary to concentrate on remembering the twelve digit number or four word keys for the current year.

Try these for practice:

September 12, 2010	_________	November 25, 2015	_________
January 17, 2016	_________	April 26, 2011	_________
February 14, 2014	_________	June 30, 2012	_________

To calculate the first Sunday of each month, each year, refer to Chapter 16 – Perpetual Memory Calendar.

Calendar Showing Date of First Sunday of Each Month

	JAN	FEB	MAR	APR	MAY	JUN	JUL	AUG	SEP	OCT	NOV	DEC
2010	3	7	7	4	2	6	4	1	5	3	7	5
	MY CAKE			RAIN HASH			RED LIGHT			MICHAEL		
2011	2	6	6	3	1	5	3	7	4	2	6	4
	NACHO CHUM			METAL			MUGGER			NATURE		
2012	7	4	4	1	6	3	1	5	2	7	4	2
	KEY ROAR			TOE JAM			TALL NICKERS			CORN		
2013	6	3	3	7	5	2	7	4	1	6	3	1
	JEMIMA			CLOWN			CART			CHIMED		
2014	5	2	2	6	4	1	6	3	7	5	2	7
	LINEN			CHART			SHOEMAKER			LINK		
2015	4	1	1	5	3	7	5	2	6	4	1	6
	WROTE TO			LAME KID			LAUNCHED			RATE CHART		
2016	2	6	6	3	1	5	3	7	4	2	6	4
	NO JUDGE			MOTEL			MY CAR			NO SHOW RAT		
2017	1	5	5	2	7	4	2	6	3	1	5	3
	THE LILY			NO CAR			TOUCH ME			THE LIME		
2018	7	4	4	1	6	3	1	5	2	7	4	2
	CAR RADIO			TOE JAM			THE LION			CAR NEWS		

12

> "The one who weaves his experiences into systematic relations with each other will be the one with the best memory."
>
> — William James

THE SENTENCE METHOD

SENTENCING is another way to remember certain types of information. In developing SENTENCES, you will probably have to do some substitution of the words representing the facts you are trying to remember for words that are more logical to the ideas of your SENTENCE. Often you will make several SENTENCES to contain all the information you wish to remember. You must know your material well before you try to SENTENCE it. Make certain that you can remember your SENTENCE in the proper order. This is generally much easier for you when you make your own SENTENCE, rather than using one that someone else has constructed. The disadvantage of the SENTENCE method is that for some material, you will only learn it in sequence, and therefore will not be able to recall it out of sequence. Of course, the KEY METHOD overcomes this disadvantage and if you know the material by both methods, then that will reinforce your ability to recall.

THE NOTES OF THE MUSICAL STAFF

In the Treble Clef:

Spaces spell FACE

Lines sentenced: Every Good Boy Does Fine.

In the Base Clef:

Spaces sentenced: All Cows Eat Grass.

Lines sentenced: Good Boys Do Fine Always.

<u>IN TRIGONOMETRY</u>

SINE	COSINE	TANGENT
<u>O</u>	<u>A</u>	<u>O</u>
H	H	A
<u>Oscar</u>	<u>And</u>	<u>Oswald</u>
Have	Had	Algebra

Here is a list of the surnames of the Heavyweight Boxing Champions, from Sullivan to Clay, in order:

SULLIVAN	DEMPSEY	CHARLES
CORBETT	TUNNEY	WALCOTT
FITZSIMMONS	SCHMELING	MARCIANO
JEFFRIES	SHARKEY	PATTERSON
HART	CARNERA	JOHANSSON
BURNS	BAER	PATTERSON
JOHNSON	BRADDOCK	LISTON
WILLARD	LOUIS	CLAY

Here is a paragraph of SENTENCES for these men's names:

When SULLIVAN opened the CARBonate the FIZZ gave JEFF HEART BURNS. JOHN WILL come out of the DAMP TUNNEL SMELLING like a SHARK. A CARNival BEAR drove a BREAD trUCK in St. LOUIS near where CHARLES sells WALNUT and MARASCHINO cherry PATTies. JOE'S HANDSOME brother PATTed his feet as he LISTENed to this table of the CLAY champs.

Notice this next one is only one sentence:

SILLY CORBETT FIGHTS JEFF'S HEART BURN, JOINING WILLARD in a DAMP TUNNEL, SMELLING of SHARKS, CANNED BEER and HADDOCK as LOUIS XIV CHAIRS WALK and MARCH PAST JOHN'S PARTY LIST of CLAY.

On a separate piece of paper make your own SENTENCE or SENTENCES for these 23 men. Let your imagination run wild. You might do it in one SENTENCE but you might take six or eight. This should take you about 10 to 20 minutes.

Here is an example of SENTENCING the Presidents of the United States in order. Note that there are some words that do not translate into the Presidents names but that add meaning to the SENTENCE. These are not at all confusing, but instead are usually necessary.

The WASHINGTON AD-MAN JEFF, got very MAD about the MONEY being spent. So the AD MAN told JACK to BURN the HAIR off the TILE and then POKE at it. His TAILOR FILLED and PIERCED the BUCKSKIN LINK belt. JOHN GRANTED that the HAZY FIELD was good for ARTISTS, especially CLEVER ones. HARRY went to CLEVELAND to KINDLE the REVOLT against TOUGH WILLIS, HARD and COOL. WHOEVER ROSE the TRUE HOURS should KINDLY tell JOHNSON, "NIX ON the CARTER RAY GUN." BUSH 1 CLEANS the TOWN of BUSH 2 in ALABAMA.

This is a masterpiece of SENTENCING because it does not have any words that are not translated.

Washington And Jefferson Made Mon Row, And Jack's Son, Van Buren Harassed Tyler, Polked Taylor. Fillmore Pierced Buchanan. "Lincoln", Johnson Grunted, "Haze Garfield". Arthur Cleverly Harassed Cleveland. Mac Kindly Rose to belt Taft. Wilson Hardly Cooled. Whoever rose to belt Truman, Heisted an hour while Kennedy Joined Nixon to find a for Carter Ray Gun to blast BUSH 1 and CLEAN the TOWN of BUSH 2 in ALABAMA.

If you want to remember the Presidents by the SENTENCE method you will be better off spending 20 minutes making your own SENTENCE rather than learning these, although you can learn these if you wish.

If you want to learn the Presidents in and out of sequence then we use the number key and linking methods. If you want to remember the year of inauguration, then we would add the number alphabet system. The chart at the end of the next chapter shows how you can add the various methods.

POEMS AS A MEMORY METHOD

POEMS can serve a purpose similar to SENTENCES and ACRONYMS for remembering certain facts or information. Of course, you have to use your ingenuity, creativity, and imagination. This is fun and very rewarding. Try it, you might discover a hidden talent.

THE DAYS OF EACH MONTH OF THE YEAR -This is probably the most universally used POEM for remembering complex information.

Thirty days has September
April, June, and November
All the rest have thirty-one
Except when February is done
It has but twenty-eight, that's fine
Till Leap Year gives it twenty-nine.

The CRANIAL NERVES, IN PROPER ORDER. This combination of a POEM and ACRONYM has helped medical students, for many years.

The cranial nerves are Olfactory, Optic, Oculomotor, Trochlear, Trigeminal, Abducent, Facial, Auditory, Glossopharyngeal, Pneumogastric, Spinal accessory, and Hypoglossal. Now see this easy little ditty:

On Old Olympus' Towering Top
A Finn And German Picked Some Hops

You might also put this into a POEM something like this example:

The Heavyweight Champs -reached the top of the ramps.
We'll learn them because we now care
The men in this list -lived by swinging their fists
Not by reading, of that we're aware.

When Sullivan fought Corbett -he was knocked out of orbit
  Then Fitzsimmons joined in the affair.
Then Jeffries, then Hart -Burns and Johnson took part
  But Jess Willard then gave them a scare.

Came Dempsey the mauler -but Tunney was taller
  Max Schmeling got smacked in the ring
Too, Sharkey was there ~ and Carnera and Baer
  And Braddock, the Cinderella King.

Joe Louis was best -and he quit for a rest
  That gave Charles, then Walcott a chance
Marciano came next -his opponents were vexed
  He retires, so Patterson could dance.

Next Johansson, the Swede -he was handsome indeed
  But Patterson got back to him too
Liston is next I say -let's all shout Hooray
  For the greatest of all, Cassius Clay.

On the Piano:

All the G and A keys
  Are between the black threes,
And tween the two's are all the D's.
  Then on the right side of the three's
Will be found the B's and C's.
  But on the left side of the three's
Are all the F's and all the E's.

13

> "I have never met a famous man or woman, and I've known most of them in my time as a newspaper writer, that did not have an absolutely fabulous memory."
>
> —Adele Rogers St. Johns

SUBSTITUTION

Since we have difficulty in remembering things that we cannot visualize we must then change the abstract or intangible into something we can visualize. This is done by substitution. If you see a word or phrase that seems abstract then substitute one that reminds you of the same thing but can be pictured in your mind.

One thing that is important to know is that if one of the substitute words we use in this course does not remind you of the thing you want to remember, then use one of your own. Use your own imagination. You must have clear pictures formed in your mind.

As an exercise in using this method, we can learn the United States and their capitals.

STATE	ASSOCIATION	CAPITAL
ALABAMA	Ali-Baba rode into Montgomery Wards.	Montgomery
ALASKA	I'll ask her in June, no?	Juneau
ARIZONA	The air is gonna cost a fee.	Phoenix
ARKANSAS	The ark can saw a little rock	Little Rock
CALIFORNIA	Cal, if you're in the corn hand me a sack of pimentos.	Sacramento

COLORADO	Color an olden fur.	Denver
CONNECTICUT	Connect your heart to a Ford.	Hartford
DELAWARE	Della wears a dover mink	Dover
FLORIDA	Flora does live in a tall glassy house	Tallahassee
GEORGIA	George is carrying a lantern.	Atlanta
HAWAII	My Hawaiian honey is a lulu.	Honolulu
IDAHO	I'd take a hoe so the boys can see me.	Boise
ILLINOIS	Ill noises spring down the field.	Springfield
INDIANA	In Diana, I see an Indian and an apple.	Indianapolis
IOWA	I own de mines.	De Moines
KANSAS	Can this toe peek at me?	Topeka
KENTUCKY	Ken took a frankfurter.	Frankfort
LOUISIANA	Louise and Ana have a bet on rouge.	Baton Rouge
MAINE	Main street was hit by a gust of wind.	Augusta
MARYLAND	A merry land is an apple place.	Annapolis
MASSACHUSETTS	Massa chews a Boston cream pie.	Boston
MICHIGAN	Mich's gun makes our land sing.	Lansing
MINNESOTA	Minnie sorta loves Paul.	St. Paul
MISSISSIPPI	Mrs. Pippy plays jacks.	Jackson
MISSOURI	Miss Sue reads in Jeffers sitting room.	Jefferson City
MONTANA	Mont and Ana are going to hell in a basket.	Helena

NEBRASKA	A new brass lincoln is my choice.	Lincoln
NEVADA	I never had a car in the city.	Carson City
NEW HAMPSHIRE	My new hamster was conked with a cord.	Concord
NEW JERSEY	My new jersey tent weighs a ton.	Trenton
NEW MEXICO	New Mexico makes Santa faint.	Santa Fe
NEW YORK	New corks are all banned.	Albany
NORTH CAROLINA	The north car lines up for the rally.	Raleigh
NORTH DAKOTA	No Dakota has a sho-biz mark.	Bismarck
OHIO	"O-HI-O great queen," said Columbus.	Columbus
OKLAHOMA	Oakla's home is in the city.	Oklahoma City
OREGON	The ore is gone. Salem took it.	Salem
PENNSYLVANIA	The pencils came from Harrah's bird.	Harrisburg
RHODE ISLAND	We rowed to the island while Provi danced.	Providence
SOUTH CAROLINA	So Carol and Ina acted at Columbia studios.	Columbia
SOUTH DAKOTA	So da coat belongs to Pierre.	Pierre
TENNESSEE	Ten is seating the bashful boy.	Nashville
TEXAS	Tex has an Austin car.	Austin
UTAH	You taught in Salt Lake City.	Salt Lake City
VERMONT	Vern and Mont, peel yer apple.	Montpelier
VIRGINIA	Virginia chases a rich man.	Richmond

WEST VIRGINIA	Wes, Virginia dances the charleston.	Charleston
WASHINGTON	I'm washing tons of clothes in Olympia beer.	Olympia
WISCONSIN	Whiskey will make you mad, son.	Madison
WYOMING	Why, O Ming, are you chasing shy Ann?	Cheyenne

NAME			SUBSTITUTION	KEY	ASSOCIATION	YEAR	WORD	ASSOCIATION
1.	George	Washington	Washington	HAT		1789	FIB	
2.	John	Adams	Adam	HEN		1797	BIKE	
3.	Thomas	Jefferson	Jeff	HAM		1801	SODA	
4.	James	Madison	Mad's Son	HARE		1809	WASP	
5.	James	Monroe	Money	HALL		1817	DIG	
6.	John Q.	Adams	Ad-Man	HASH		1825	NILE	
7.	Andrew	Jackson	Jack's Son	HOOK		1829	KNOB	
8.	M. Van	Buren	Burn	HOOF		1837	MUG	
9.	Wm. H.	Harrison	Harry's Son	HOOP		1841	ROOT	
10.	John	Tyler	Tile	TOES		1841	ROAD	
11.	James	Polk	Poker	TOT		1845	ROYAL	
12.	Zachary	Taylor	Tailor	TON		1849	RIP	
13.	Millard	Fillmore	Fill More	TIME		1850	WHEELS	
14.	F.	Pierce	Pierced	TAR		1853	LIMB	
15.	James	Buchanan	Buckskin	TAIL		1857	LEG	
16.	Abraham	Lincoln	Lincoln	TISSUE		1861	SHOT	
17.	Andrew	Johnson	John's Son	TACK		1865	JELLY	
18.	U.S.	Grant	Grant	TAFFY		1869	SHEEP	
19.	R.	Hayes	Hazy	TAPE		1877	COKE	
20.	James	Garfield	Garlic	NOOSE		1881	FOOD	
21.	Chester	Arthur	Artist	NET		1881	PHOTO	
22.	Grover	Cleveland	Clever	NUN		1885	FLY	
23.	Ben	Harrison	Harry's Son	NAME		1889	V.P.	
24.	Grover	Cleveland	Cleve	NERO		1893	BEAM	
25.	Wm.	McKinley	Mac Kindly	NAIL		1897	PACK	
26.	Theodore	Roosevelt	Revolting	NOTCH		1901	SUIT	
27.	Wm. H.	Taft	Tough	NOOK		1909	SOUP	
28.	Woodrow	Wilson	Willow	KNIFE		1913	DAM	
29.	Warren	Harding	Hard	NAP		1921	NIGHT	
30.	Calvin	Coolidge	Cool	MESS		1923	NUMB	
31.	Herbert	Hoover	Whoever	MAT		1929	NIP	
32.	F.D.	Roosevelt	Rose's Belt	MOON		1933	MAMA	
33.	Harry	Truman	True Man	MUMMY		1945	REEL	
34.	Dwight	Eisenhower	Eyes on the Hour	MARE		1953	LAMB	
35.	John	Kennedy	Kindly	MAIL		1961	SHADE	
36.	L.B.	Johnson	Johnson	MATCH		1963	GYM	
37.	Richard	Nixon	St. Nick	MIKE		1969	SHOP	
38.	Gerald	Ford	Ford	MUFF		1974	CHOIR	
39.	James	Carter	Cart	MOP		1977	KEG	
40.	Ronald	Reagan	Raygun	RACE		1981	FOOT	
41.	Geo	Bush	Bush 1	RAT		1989	FOB	
42.	**Wm**	**Clinton**	**Clean Town**	**RAIN**		**1993**	**PALM**	
43.	**Geo W.**	**Bush**	**Bush 2**	**RAM**		**2001**	**SEAT**	
44.	Barak	Obama	Obama	ROWER		2009	SOAR	

14

> When asked, "Can I use my memory as effectively as you?" Dr. Joyce Brothers answered, "Certainly – absolutely – without doubt."

ACRONYMS

These are compositions in which certain letters are taken in or out of order to form a word, which can be a title, name, or motto, that can be associated with the subject of the composition. Often we take the first letter (or the first few letters) of several words to make a word. We use many of these everyday. Probably the most frequently used ACRONYM, is the word NEWS, which means, information from North, East, West and South. Another example is the word RADAR, which means, RAdio Detecting And Ranging. Government bureaus use these quite extensively, such as NATO, UNESCO and NAVBUPERS.

ACRONYMS have a definite place as a memory method. They can hold together facts you wish to remember for many years with very little review because of the novelty of the word or words you use. Of course, you must know the information very well before you would use the ACRONYM. Some people think of these as a sort of One Eyed Method, but we must consider that, in the Land of the Blind, the One Eyed Man is King. The best ACRONYMS are the ones that you make yourself for facts you wish to keep in mind. They are excellent for teaching or lecturing when you want the audience to remember your most important points.

In making your ACRONYM, use scratch paper and write the letters you can use from the facts you are going to remember. Decide whether or not these facts have to be in order, for example, the RULES FOR REMEMBERING PEOPLE'S NAMES should be in order, while the RULES OF LISTENING need not be. Now scc if you have more than one choice of letter for one point, for instance, RULE 1 OF NAMES, GET THE NAME CORRECTLY, you have these choices:

G (Get It)
C (Correctly Get It)
T (The Name Correctly)
N (Name Correctly)

You should try to get several words, or combination of words, before you decide on the one you will use. A dictionary is very helpful in making ACRONYMS. Playing games, such as .Scrabble or Scribbage are excellent practice in making ACRONYMS, as well as in building your vocabulary. Here are some examples of ACRONYMS for you to study, however you might find better ones yourself.

RULES FOR REMEMBERING PEOPLE'S NAMES — <u>G.I.-A-OK</u>

G et it
I mmediately repeat it
A ssociate it
O bserve features
K eep in mind

5 MOST IMPORTANT RULES OF LISTENING — WET AD

W ork at listening
E ye contact
T hink at speaker's speed
A sk questions
D on't jump to conclusions

THREE IMPORTANT FACTORS IN MEMORY — CAR

C oncentrate
A ssociate
R eview

THE GREAT LAKES — HOMES (They are initialed on your map)

H uron
O ntario
M ichigan
E rie
S uperior

HORSEPOWER — That's a PLAN

P ressure
L ength
A rea
N umber of revs

COLORS OF THE SPECTRUM, IN ORDER — ROY G. BIV

R ed
O range
Y ellow
G reen
B lue
I ndigo
V iolet

From the Wonderful Book and Recording PSYCHO-CYBERNETICS, by MAXWELL MALTZ, M.D. — SUCCESS

S ense of Direction
U nderstanding
C ourage
C harity
E steem
S elf-Confidence
S elf-Acceptance

THE 5 RULES OF ADVERTISING — AIDCA

A ttention
I nterest
D esire
C onviction
A ction

THE PRESIDENT'S CABINET — DIC ASPHALT

D efense
I nterior
C ommerce
A ttorney General
S tate
P ostmaster General
H ealth Education & Welfare
A griculture
L abor
T reasury

(Other possibilities are ST. HADICLAP or DIALS PATCH or TAPS A CHILD or SAD HAT CLIP or SLID A PATCH and on and on and on.)

A professional man, upon leaving the plane, said to the hostess. "Thank you for an enjoyable flight, Miss Miller."

His wife, upon overhearing the remark stated, "Bill Lindon, why are you always so interested in the girls?"

"Don't be so suspicious, honey, you know that the pilot, copilot, and hostesses names are inside of the door of the plane."

"Okay," said the fuming wife, "What are the names of the pilot and copilot?"

15

> "If a man empties his purse into his head, no man can take it away from him. An investment in knowledge always pays the best interest."
>
> — Benjamin Franklin.

PLAYING CARDS

How long would it take you to memorize a deck of playing cards in sequence? Perhaps you think such a feat is impossible, or extremely difficult.

Before you finish this chapter you will find how easy it is to memorize a deck of cards in sequence in less than ten minutes. YOU will also amaze your friends by performing other impressive memory feats with playing cards.

This section on playing cards is included here only as a -memory aid. Cards, as you will use them, are excellent for improving associations for quickly linking thoughts and ideas together. Practice with the cards will help you thoroughly learn some of the Number Keys.

Please bear in mind that this chapter does not teach you card games, but is included only to help you use playing cards as a memory exercise.

TRANSLATING CARDS INTO NUMBER KEY WORDS

Each playing card must be translated into one of the Number Keys. This is important, because, just like numbers, cards have no meaning for us. Therefore, we must make them meaningful.

We translate cards into numbers as shown in the following chart:

SUIT	CARD	NUMBERS	NUMBER KEYS
CLUBS,	Ace to King,	21-33	Net to Mummy
DIAMONDS	Ace to King	41-53	Rat to Lime
HEARTS	Ace to King	61-73	Jet to Comb
SPADES	Ace to King	81-93	Fight to Palm

You will note that the suits are in alphabetical order, or the same order of value as in many card games where Clubs have the lowest suit value.

EXAMPLES:	Ace of Clubs	Equals	Net
	Eight of Clubs	Equals	Knife
	Ten of Clubs	Equals	Mess
	Jack of Diamonds	Equals	Light
	Queen of Hearts	Equals	Cane
	King of Spades	Equals	Palm
	Joker	equals	Joker

MEMORIZING A DECK OF CARDS IN SEQUENCE

To memorize a deck of cards in sequence, you simply link the Card Key words together to form a chain. Be sure to start the chain with a known Key such as hat or an item in the room – a desk or a table. You must use a known Key in order to know where you started your chain. If you only linked the fifty-two cards together you might have difficulty in determining which card was the first one in the deck. Be sure to bring the last card in the deck back and link it with your starting Key. In this manner, your chain is complete.

EXAMPLE:

Key	6C	2C	6D	QS	AH	9S	5C
hat	notch	nun	roach	pin	jet	fob	Nail
JD	8D	8C	KD	3H	KC	Key	
light	roof	knife	lime	jam	mummy	hat	

The example above is one-fourth of a deck of cards. With a little practice you can memorize a full deck of cards in ten minutes or less. Some students are able to memorize a deck of cards in less than three and one-half minutes. Practice is the key to improving your speed in memorizing a deck of cards. The more you practice linking cards together, the more proficient you become in linking other thoughts, ideas, and objects. Thus, the more valuable this memory training will be to you.

One major advantage in using a deck of cards in this phase of your memory training is that when you shuffle the deck, you have a whole new set of ideas to use for practice.

THE MISSING CARD EXPERIMENT

Have someone shuffle a deck of cards and then cut the deck. Now, have him slowly read about one-half of the deck to you. When he has done this you will be able to name all the cards he did not read to you, or the missing cards.

To do this experiment, you form an association between yourself and the Key Word of the card read. For example, if the Ace of Clubs is read, you picture yourself tangled in a net, and so on, for every card read. To recall the missing cards, you start with the Ace of Clubs and mentally review the deck asking yourself, "Was I doing something with the key word for each card?" If you have an association for a card, then that card was read to you. If you find that you do not have an association, then that card is one of the missing cards. You continue in this manner until you have completed the entire deck.

A popular way to use the missing card experiment as a memory demonstration is to have someone deal a hand of Bridge. Then have three persons read the thirteen cards in their hands, and you name the cards in the fourth person's hand.

TRICK MIND READING

In order to do this trick, you will need one other person familiar with the above playing-card experiments. First, you must have the deck memorized, as explained earlier in this chapter, or you may wish to arrange them by the following method: Just learn this sentence:

Eight	Kings	ThreaTened		To	Save	Ninety-Five	
8	K	3	10	2	7	9	5
	Queens	From		A	Sick	Jack	
	Q	4		A	6	J	

This sentence gives you each card from Ace to King one time, so by repeating it four times you have arranged a full deck. By alternating the suits; Spades, Hearts, Diamonds, and Clubs you will have a deck arranged in the following manner: 8S KH 3D 10C 2S 7H 9D 5C QS 4H AD 6C JS 8H KD, etc.

You may cut the deck as many times as you like and you will not affect the sequence. This arrangement mixes the deck to the extent that a person who is unfamiliar with this method, will actually think it has been shuffled.

The person with the deck can go out into an audience and let someone draw a card by fanning the deck from one hand to another, face down. When a card has been selected, he places all the cards above the selected card on the bottom of the deck. He then nonchalantly looks at the bottom card. By so doing, he can tell which card has been taken from the deck. He then repeats a phrase or sentence. The first letter of the first word represents the suit. The first letter of the second word represents the card value. Any other words in the sentence have no meaning to the "mind reader."

The "mind reader," who can be blindfolded, then announces to the audience which card was taken from the deck. Then the "sender" takes that card back, cuts the deck a few times, and proceeds to another person and the trick is repeated.

To expand the number of possible words for use in this experiment — AND IN THIS EXPERIMENT ONLY — we use the W as a 3 and H as an 8. Of course, we continue to use M as 3 and F or V as 8.

EXAMPLES:

AC — Now Try	2H — Just Now
5C — Never Let	6H — Just Choose
JC — What Do	QH — Concentrate Now
KC — Well What	4S — Here, Read
3D — Ready, What	8S — How Fine
9D — Read Please	10S — Please Say

This same principle can be used also for naming the date on a coin, or any other two-digit number.

EXAMPLES:

1929 — New Please	1961 — Just Try
1885 — Here Let's	1904 — Say, Read

Of course, the secret of success with these and all memory feats, is practice. The more you practice, the more rapidly you will be able to perform these demonstrations.

16

> "An alert memory will give you a shortcut to success and personal achievement."
>
> — Charles M. Simmons

PERPETUAL MENTAL CALENDAR

Would you like to keep in mind the day of the week for every date in the current year? If you are one who must refer to a calendar often, then you will find the PERPETUAL MENTAL CALENDAR a practical time saving device Once you have learned to use the PERPETUAL MENTAL CALENDAR you can mentally calculate dates faster than you can refer to a desk or pocket calendar.

The calendar in use today is known as the Gregorian Calendar. It was designed by Clavious, an astronomer, and ordained by Pope Gregory XIII, on October 15, 1582 (elephant).* This calendar was adopted by an English statute in 1752 (calendar),* and is now used by all countries. The last country, Greece, adopted it in 1923. Earlier calendars were inaccurate because of the difficulty in dividing the true year of 365.242 (mutually unarranged)* days, into an even number of days. The Gregorian Calendar is so accurate that it will take 3,261 (whim unwished)* years to gain one day.

*Word into number translation to help you remember the number.

HOW TO USE THE PERPETUAL MENTAL CALENDAR

FIRST: We must determine the value of the year that we want to remember by the following formula:

> Century plus Score plus Unit minus Highest multiple of seven equals the Year Value. The formula in short:

C+S+U-H=Y

We now fill into this formula, numbers from the following tables:

CENTURY	SCORE	UNITS				
(100 Years)	(20 Years)	(Number of years past the score)				
1600=6	0=0	0=0	4=5	8=10	12=15	16=20
1700=4	20=4	1=1	5=6	9=11	13=16	17=21
1800=2	40=1	2=2	6=7	10=12	14=17	18=22
1900=0	60=5	3=3	7=8	11=13	15=18	19=23
2000=6	80=2					

HIGHEST MULTIPLE OF SEVEN

This is the highest of the following numbers that can be subtracted from the sum of the C+S+U. 0-7-14-21-28-35-42.

EXAMPLES:

1809	2+0+11-7 = 6	1962	0+5+2-7 = 0
1776	4+5+20-28 = 1	2027	6+4+8-14 = 4

It is not necessary for you to memorize this formula or the numbers in the tables, but you should be familiar with it so that on the first of each year you can calculate the Year value and keep it in mind for the rest of the year. Because of the Highest Multiple of 7, the Year value can only be a number from 0 to 6. Now that you understand this you are ready for the next step.

SECOND: Since Y easily can be remembered for the current year, it is a simple task to know the day of the week for any date in the year.

To determine the day of the week, use the following formula:

Year plus Month plus Number of the date minus the Highest multiple of seven equals Day of the week. The formula is written:

Y+M+N-H=D

In order to remember the number value of each Month, we use the major holiday in that month for our word-to-number association. We translate the first letter of the name of the holiday association word into the number of the number alphabet equaling the value of the month.

Of course, if you can think of a better word than the examples given, use it:

MONTH VALUE CHART

MONTH	HOLIDAY	ASSOCIATION	VALUE
January	New Year's Day	Time	1
February	Valentine's Day	Remembrance	4
March	St. Patrick's Day	Ireland	4
April	Easter	Sunday	0
May	Mother's Day	Nice	2
June	Father's Day	Loaf	5
July	Independence Day	Scorcher	0
August	Vacation Time	Mountains	3

September	Labor Day	Jobs	6
October	Halloween	Tots	1
November	Thanksgiving	Reverent	4
December	Christmas	Jesus	6

The Number of the date is the date you are interested in using.
The High multiple of seven is the same as in the previous formula.
The Day of week is just as it is on the calendar:

1	Sunday
2	Monday
3	Tuesday
4	Wednesday
5	Thursday
6	Friday
0	Saturday

EXAMPLES:
Let's use examples utilizing the years we found the value of as above:

February 19, 1809

1809 equals 6
February equals 4
February 19 equals 19
High multiple of seven is 28

So, we have:

6 plus 4 plus 19 minus 28 equals 1 which is Sunday.

July 4, 1776

1 plus 0 plus 4 minus 0 equals 5 which is Thursday.

August 23, 1962

0 plus 3 plus 23 minus 21 equals 5 which is Thursday.

December 25, 2027

4 plus 6 plus 25 minus 35 equals 0 which is Saturday.

LEAP YEARS

Leap years are the years divisible by four except for the century years, which are leap years only every fourth century beginning in 1600. For instance, 1600 is a leap year, as is 2000. However, 1700, 1800, 1900, 2100 are not leap years. IMPORTANT: In January and February of leap years, it is necessary to deduct one from the day which means you go back one day.

You will notice as you use the formula for each year, the year value advances only one number, instead of on leap year when it advances two numbers. Thus, instead of figuring the year value by the formula each year, just add one number (or two on leap year) to the value of the previous year. That is, of course, until you reach the number six, at which time you start over again the next year with zero.

If we know the day of the week a certain thing is to happen, we can determine the date by the following method:

1st week	starts	on the	1st
2nd week	starts	on the	8th
3rd week	starts	on the	15th
4th week	starts	on the	22nd
5th week	starts	on the	29th

EXAMPLE:

You want to know what date the third Wednesday will fall on in December, 1962. Determine what day the fifteenth of December will be as follows:

0 + 6 + 15 -21 = 0 which is Saturday.

From Saturday to Wednesday is four days.

Thus: 4 + 15 = 19.

So the date you want falls on the 19th of December.

The more you use and practice the Mental Calendar, the more skilled you will become.

17

> "Your memory -speedway to success in earning, learning and living."
>
> — O.W. Bill Hayes

HOW TO REMEMBER GRAPHICS

You learn this method by using your Map. Notice that the east to west lines are slightly curved, because the map is a Polyconic Projection. On a Mercator projection these are straight, but for our purpose this makes little difference. You are going to draw the outside boundary of this map from memory, on a separate paper, so you must study the boundary in the following manner:

Note the reference numbers; 1 is the niche in north west Washington; 2 is Eureka; 3 is San Diego; 4 is near Langdon; 5 is Hastings; 6 is near Lawton; 7 is McAllen; 8 is Van Buren; 9 is New York City; 10 is Charleston, and 11 is Key Largo. Look at the lines 2 to 9, 3 to 10, also 4 to 7 which divides 2 to 9 and 3 to 10 in half.

Follow the text in detail and understand each part before progressing, while looking at your Map. Study the border starting at 1; see the line go past 4 to the up jog at Lake of the Woods, Minn. (the most northern point) — notice the curved line through Lake Superior and Lake Huron down to the southern part of Canada in Lake Erie just south of line 2 to 9 — study the irregular line through Lake Erie and Lake Ontario and up the St. Lawrence River, then east — then north — then east (a lazy S shape) to 8 — note that the northern part of Maine is south of the 1 to 4 line continued in its curve — see the short line from 8 to the eastern most point of the U.S. at the Atlantic Ocean — now look at the entire eastern coast, the curved line from Maine to Cape Cod (the point just east of Boston) is reproduced at twice the length from Cape Cod to Cape Hatteras (the point east of Raleigh), the curved line from Cape Hatteras to Key Largo is similar to the other two, and is equal in length to Maine to Cape Hatteras — notice the approximately 30° westward slope from Main to Key Largo — from 11 to the southern tip of Texas you see a semi-circle somewhat irregularly flattened on the upper side — note that the tip of Texas is just a little west of Lake of the Woods — the southern border of Texas (along the Rio Grande) passes 7, the curve duplicates itself upside down and then again right side up to El Paso (at the New Mexico border) — then study the series of 5 straight lines that take you to Sari Diego at 3 — notice the 45° angle north west to San Francisco (the Bay nearest Sacramento) — look north from San Diego and you will see that about 1/3 of Nevada and 9/10 of Washington and Oregon are to the west --the line from San Francisco curves northward and at 2 it straightens out and goes north — finish your niche at 1.

You have studied the boundary. Put the printed map out of sight now and draw it. Take a clean piece of paper and pencil with eraser and thinking of what *you* studied, start at 1 and go all the way around. Try your best to make your angles and curves in correct shape and proportion. Now number the 11 points and draw the 3 lines. Erase and redraw any parts necessary. Now compare it to the printed Map and look for any mistakes (they will really stand out), and study them. Within a week, draw another map from memory and compare it. This second one should be quite accurate. When *you* do this 2, 3 or 4 times *you* will have it in your memory forever, with considerable accuracy.

Since you already know the States and capitals, and also where they are on the Map, why not really step out and draw them in on your last map, from memory?

The principle is that, in drawing from memory, *you* must know the graphic so well that *you* will remember it. This method can be applied to remembering Maps, Charts, Graphs, Mechanical Parts, Plant and Animal Anatomy or anything else that can be seen.

-Hum, was Louisiana a Boot or a sandal?

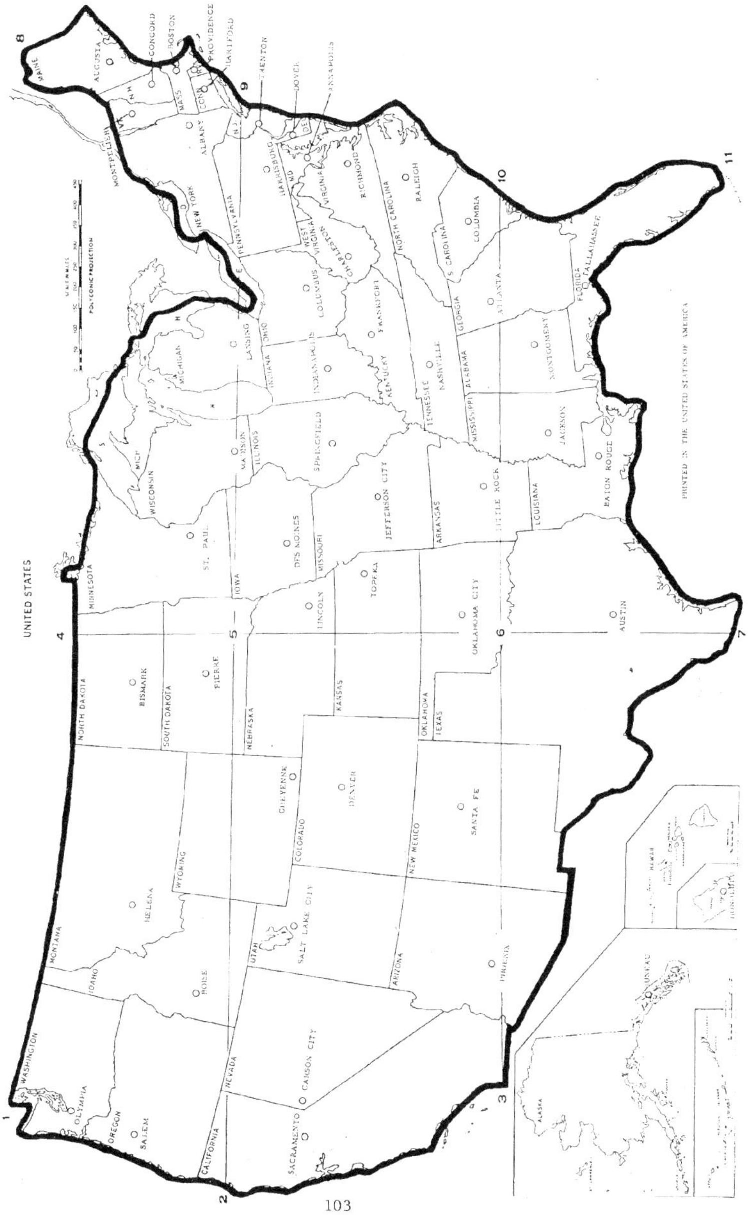
UNITED STATES
POLYCONIC PROJECTION
1
2
3
4
5
6
7
8
9
10
11
WASHINGTON
OLYMPIA
OREGON
SALEM
CALIFORNIA
SACRAMENTO
NEVADA
CARSON CITY
IDAHO
BOISE
MONTANA
HELENA
WYOMING
CHEYENNE
UTAH
SALT LAKE CITY
ARIZONA
PHOENIX
COLORADO
DENVER
NEW MEXICO
SANTA FE
NORTH DAKOTA
BISMARK
SOUTH DAKOTA
PIERRE
NEBRASKA
LINCOLN
KANSAS
TOPEKA
OKLAHOMA
OKLAHOMA CITY
TEXAS
AUSTIN
MINNESOTA
ST. PAUL
IOWA
DES MOINES
MISSOURI
JEFFERSON CITY
ARKANSAS
LITTLE ROCK
LOUISIANA
BATON ROUGE
WISCONSIN
MADISON
ILLINOIS
SPRINGFIELD
MICHIGAN
LANSING
INDIANA
INDIANAPOLIS
OHIO
COLUMBUS
KENTUCKY
FRANKFORT
TENNESSEE
NASHVILLE
MISSISSIPPI
JACKSON
ALABAMA
MONTGOMERY
GEORGIA
ATLANTA
FLORIDA
TALLAHASSEE
S CAROLINA
COLUMBIA
NORTH CAROLINA
RALEIGH
VIRGINIA
RICHMOND
WEST VIRGINIA
CHARLESTON
PENNSYLVANIA
HARRISBURG
NEW YORK
ALBANY
MAINE
AUGUSTA
N H
CONCORD
VT
MONTPELIER
MASS
BOSTON
CONN
HARTFORD
PROVIDENCE
N J
TRENTON
DEL
DOVER
MD
ANNAPOLIS
ALASKA
JUNEAU
HAWAII
HONOLULU
PRINTED IN THE UNITED STATES OF AMERICA

103

MAP DRAWING

Map drawing affords excellent practice for students of memory improvement because it requires close observation. I dare say not one person in thousands can draw from memory a rough outline of the United States.

Drawing an outline of the State in which he lives is usually impossible for the average person, unless he lives in Wyoming or Colorado. This inability is not at all surprising. The eye has not been trained to remember what it sees.

By using a little imagination and some regular practice you will find many interesting things in map drawing that will aid the memory.

For example, the lower portion of the State of Michigan resembles an old hunched-up woman with her hands in a shawl trying to keep warm. Also note the topknot on her head.

The outline of Italy closely resembles a high-heeled boot drawn back to kick a stone out of the way. The stone being the island of Sicily.

With a little imagination, the outline of England and Wales may resemble an old man riding a hobby horse.

The continent of South America looks like an elephant's head and trunk.

France, Spain, and Portugal bear a close resemblance to a pig's head with a snout in a bucket.

Drawing lines between the cities in the states in which you are interested will aid the mind's eye in fixing their location as well as the outline of the state.

By making these outlines a few times you will get deep impressions on the mind and readily develop a geographical memory.

18

> "That which we persist in doing becomes easier; not that the nature of the thing itself changes, but rather our ability to do increases in direct proportion to our effort and persistence."

AND SO ... THE END WHICH IS REALLY THE BEGINNING

The methods we have outlined for you in this book are designed to be your servants. So master them well. When you do, you will be able to apply them in countless situations where using these techniques of memory you will be able to remember things that neither you nor your friends ever thought would be possible. So practice your memory skills frequently and you will become more and more proficient and therefore confident of your own ability to remember.

As you use your memory, you will find that you will have improved powers of visualization, concentration and observation.

And you will enjoy not only increased mental agility, but self-confidence, success in your business and social life, and loads of fun along the way.

Study well, start unlocking the unlimited potential of your mind by improving your memory skills, and then have a Great Life enriched by a Fabulous Memory!

Don A. McKinley

APPENDIX A

<u>EASY WAY TO REMEMBER THE 10 COMMANDMENTS</u>

Found in (Ex)odus (X) chapter XX

1. One God. No other Gods before me.
(Visualize "<u>1</u>"God)

2. Thou shalt not bow down or serve any graven images.
(Visualize "2" as a person bowing down or kneeling)

3. Thou shalt not take the name of the Lord in vain.
(Not from my lips, visualize a pair of lips)

4. Thou shalt keep the sabbath day holy.
(Visualize a "4" with a hole in it)

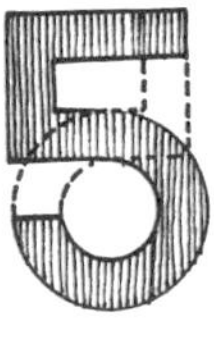

5. Thou shalt honor thy father and thy mother.
(Visualize dad as a square, mother as a perfect circle)

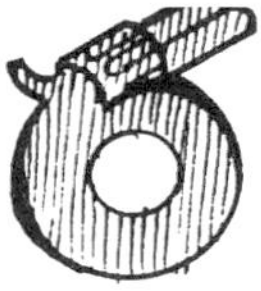

6. Thou shalt not kill.
(Visualize a six gun as a "6")

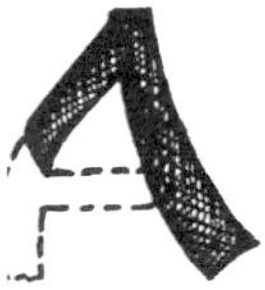

7. Thou shalt not commit adultry.
(Visualize 7 as an "A")

8. Thou shalt not steal.
(Visualize steel handcuffs as an "8")

9. Thou shalt not "bear" false witness.
(Visualize the 9 as a bear)

10. Thou shalt not covet.
(Visualize "10" as co (vet)

APPENDIX B

NUMBER ALPHABET DICTIONARY

OO	SAUCE, zoos, seesaw, hoses, oasis, icehouse
01	SEAT, suit, seed, city, acid, stew, soda, soot
02	SUN, sign, zone, snow, swan, swine
03	SEAM, Siam, swim, asthma
04	SOAR, sour, Czar, sewer, Sarah, sore
05	SAIL, seal, soil, cell, sleigh, easel, sly, sale
06	SASH, siege, sachet, switch, sage
07	SACK, sock, ski, squaw, whiskey, sick, sky
08	SAFE, sofa, salve, sieve, housewife
09	SOAP, soup, sap, sub, subway, spy, wasp
000	SEESAWS, seizes
001	SISTER, cyst
002	ASSASSIN, season, Susan
003	SEISMOGRAPH, sesame
004	SCISSORS, Ceasar, saucer
005	SIZZLE, Sicily
006	SWISS-WATCH
007	SEASICK, ice-skate
008	SISSIFIED, sassafras
009	SUSPENDERS, cesspool
010	SUDS, seeds, seats, sawdust, citizen, hostess
011	STADIUM, student, studio, stethoscope
012	STONE, stein, stain, stunt, satin
013	STEAM, stem, stamp, stump
014	STAR, stare, steer, store
015	STOOL, steel, stallion, saddle
016	STAGE, stitch, station, stooge, citation
017	STEAK, stick, stalk, stake, stack, stuck, stag
018	STOVE, staff, Steve

019 STEEPLE, staple, step, stoop, stab, stub
020 SUNSET, signs, sneeze, zones
021 SAINT, cent, sand, sentry
022 SON-IN-LAW, synonym
023 CINEMA, cinnamon
024 SNARE, sinner, scenery
025 SNAIL, senile, sunlight
026 SUNSHINE, century, singe
027 SNAKE, sneak, snack, sink
028 SNOWFLAKE, sunflower, snuff, sniffle
029 SNOWBALL, sunburn, snapshot, sniper, snob
030 SAMSON, seams, swims
031 CEMETERY, summit, smooth, smut
032 SALMON, seaman, summons, cement
033 SWIM-MEET, semi-monthly
034 SWIMMER, summer, simmer, smear, smurf
035 SMILE, smell, small, sawmill
036 SMASH, smooch, smudge
037 SMOKE, smog, smack
038 SYMPHONY, semaphore, semi-final
039 SWAMP, sample, sympathy, sombrero
040 SOARS, sewers, series
041 SWORD, sardine, wizard
042 SIREN, syringe, serenade
043 SERMON, serum, ceramic, ceremony
044 SORORITY, soarer
045 SIRLOIN, cereal, Israel
046 SURGEON, search, sergeant
047 CIRCUS, circle, circuit
048 SURF, servant, survivor, survey
049 SYRUP, serpent, surprise
050 SAILS, seals, weasels, whistles, slice
051 SLED, slide, sleet, salt, salute, salad, slot

052 SALOON, slender, slain
053 SALAMI, slalom, slumber, slam, slum
054 SAILOR, cellar, solar, salary, celery
055 CELLULOID, slyly, slowly
056 SOLDIER, solution, slush, zoology
057 SLACKS, silk, slug
058 SLAVE, sleeve, silver
059 SLIPPER, sleep, slab, slap, slip
060 SWITCHES, sashes, seiges
061 SWITCHED, swished
062 SESSION, sea-channel
063 HOUSE-CHIME, switchman
064 SEASHORE, seizure, suture, sojourn
065 SEASHELL, satchel, social
066 WISE-JUDGE, situation
067 SWITCH-KEY, sash-curtain
068 ICE-SHOVEL, easy shave
069 SWITCH-PANEL, sagebrush, sea-ship
070 SOCKS, sex, sacks
071 SKATE, skit, scout, skid, scooter, socket
072 SCHOONER, scones, scan, skin, signal
073 ESKIMO, sockem, scheme, scamp, skim
074 CIGAR, skier, screw, scar, sugar
075 SKULL, scald, scale, school, skillet
076 SCOTCH, sketch, suction, squash
077 SQUEEK
078 SCAFFOLD, skiff, scavanger, scuffle
079 SCUBA, scoop, scab, ski-boat, psychopath, whiskbroom
080 SOFAS, safes, sieves, saves
081 SOFT, sift, safety, swift
082 SYPHON, souvenir
083 SEA-FOAM, sophomore, save-me
084 SAPPHIRE, cipher, suffer, sphere

085 ASPHALT, ice-floe, civil, swivel
086 SEAFISH, savage
087 SAFEKEEPING, civic, suffocate
088 SEA-FEVER, safe-food
089 SOFA-BED, safe-peak, safe-pony
090 SPOUSE, spies, spice, space
091 SPIDER, spade, spout, spit
092 SPOON, spine, spaniel, sponge
093 SPAM, submarine
094 SPEAR, zipper, zapper, zebra
095 SPOOL, sable, splinter
096 SPEECH, spatula, species
097 SPOOK, spike, spokes, speak, spaghetti
098 SOAPFLAKES, subway-fare
099 SOAPBOX, soap-bubbles, subpoena
0 SAW, house, hose, ice, sew, zoo, ace, sea, sow
1 HAT, hut, head, tee, tie, toe, toy, auto, weed
2 HEN, hyena, honey, knee, Noah, inn, wine
3 HAM, hem, home, hymn, Ma, May, mow
4 HARE, ear, oar, wire, arrow, row, hair, ore
5 HALL, heel, halo, holly, hill, law, oil, owl, wheel, whale, eel
6 HASH, hedge, jaw, shoe, wash, watch, witch, jay
7 HOOK, hog, hockey, hawk, oak, cow, egg, key, wig, cue
8 HOOF, hive, wife, wave, ivy
9 HOOP, hobo, whip, web, bee, boy, bow, pie, ape, pea
10 TOES, toys, oats, woods, dice, daisy
11 TOT, teeth, tide, toad, tattoo, date, diet
12 TON, tin, twine, dean, dawn, dune, iodine, down
13 TIME, team, tomb, tam, thumb, dime, dam, atom
14 TAR, tire, tower, tray, tree, deer, door, heater, water
15 TAIL, tile, tool, towel, doll, duel, hotel, dial, idol
16 TISSUE, teach, twitch, dish, ditch, dash
17 TACK, toga, twig, attic, deck, dock, duck, dog, dike

18 TAFFY, TV, thief, dive, dove
19 TAPE, tube, tub, tap, dope
20 NOOSE, knees, nose, noise, henhouse
21 NET, nut, knot, knight, nude, wind, window, hand, hound, ant
22 NUN, noon, onion, union
23 NAME, gnome, enemy
24 NERO, weiner, winery
25 NAIL, Nile, knoll, inlay, kneel
26 NOTCH, hinge, wench, nash
27 NOOK, neck, nag, hanky, ink, wink
28 KNIFE, navy, nephew
29 NAP, knob, honeybee
30 MESS, Messiah, mice, moose, moss, hams
31 MAT, maid, meat, moat, moth, mitt, mud
32 MOON, mane, mine, money, man, woman
33 MUMMY, Miami, mama, mom, mime
34 MARE, Mayor, mower, hammer
35 MAIL, meal, mile, mole, mall, mill, mule
36 MATCH, mush, mesh, image
37 MIKE, mug, hammock
38 MUFF, mafia, movie
39 MOP, map, mob, hemp, hump, imp
40 RACE, rice, rose, ears, hearse, horse, warehouse
41 RAT, raid, radio, road, rot, root, herd, heart, wart, wreath
42 RAIN, iron, urn, arena, horn
43 RAM, Rome, rum, harem, worm, army, room
44 ROWER, rear, rare, roar, horror, Warrior
45 RULE, rail, relay, reel, roll, rally, aerial
46 ROACH, rage, rich, ridge, rash, rouge, Russia, arch
47 RAKE, rack, rug, rookie, ring, rock, ark
48 ROOF, reef, reweave, wharf
49 ROPE, rib, robe, harp, ruby, rabbi
50 LACE, lease, lice, lass, lasso, wheels, walls

51 LIGHT, lathe, lady, lead, lot, eyelid, wallet
52 LION, lane, lawn, line, woolen
53 LIME, lamb, limb, 100m, elm, helm
54 LAIR, lawyer, whaler
55 LILY, lulu, loyal, whale-oil
56 LASH, leech, leash, ledge, lodge, latch
57 LOCK, lake, leak, log, leg, elk
58 LEAF, lava, loaf, elf, wolf
59 LIP, lap, lab, lobby, loop, elbow, alps
60 JUICE, cheese, chess, shoes, ashes
61 JET, jade, shade, sheet, shoot, shed, chute, shadow
62 JUNE, Jonah, chain, china, chin, shine, ocean
63 JAM, gem, gym, chime
64 JAR, jury, cherry, cheer, chair, shower, usher, washer
65 JAIL, jewel, jelly, je110, shell, shawl, chile
66 JUDGE, choo-choo, Joshua
67 JACK, jog, jug, jockey, check, chalk, chick, sheik, shag
68 JAVA, chief, chive, chevy, chef, shave
69 JEEP, job, jab, sheet, shop, ship, chip, chop
70 CASE, cows, kiss, gas, goose, hogs, wigs, wax, wicks, axe
71 CAT, cot, coat, caddie, kid, kite, gate, goat
72 CANE, can, cone, canoe, queen, gun, gown, wagon
73 COMB, cameo, game, gum, wigwam
74 CAR, choir, cry, crow, gear
75 COIL, clay, claw, coal, goal, glue, eagle, quail
76 CASH, cage, couch, coach, gauge
77 COKE, cake, cock, cook, king, keg
78 CAFE, cave, cove, cuff, calf, coffee
79 COP, cape, cup, cube, cap, cab, cob, hiccup
80 FUSE, face, hoofs, hives, vase, vise, waves, wives
81 FIGHT, fat, food, foot, photo, vat, void
82 FAN, fin, phone, van, vine, oven, heaven, vein
83 FOAM, fame, fume

84 FIRE, fur, fry, ferry, fairy, ivory
85 FILE, fly, flea, veil, veal, valley, waffle
86 FISH, fudge, voyage
87 FAKE, fork, fig, fog
88 FIFE, Fifi
89 FOB, fib, Phoebe, V.P.
90 PASS, peas, posse, pizza, pies, boys, bees, boss, bus, base
91 POT, pit, poet, bat, body, bed, boot, beet, beads, boat
92 PIN, pony, pan, pen, piano, bean, bone
93 PALM, poem, plum, bomb, bum, beam, opium
94 PIER, pear, bar, beer, bear, berry, opera
95 POOL, pole, pillow, ball, bell, bull, ballet, apple, eyeball, bowl
96 POUCH, peach, patch, badge, beach, bush
97 PEAK, pick, pack, pig, beak, bag, bike, book, bug, bouquet
98 PUFF, beef, buff, beehive
99 PIPE, pope, poppy, puppy, baby, bib
100 DAISIES, dishes, disease
101 TOAST, test, twist, dust, duster

APPENDIX C

CHILDREN'S MEMORY AID

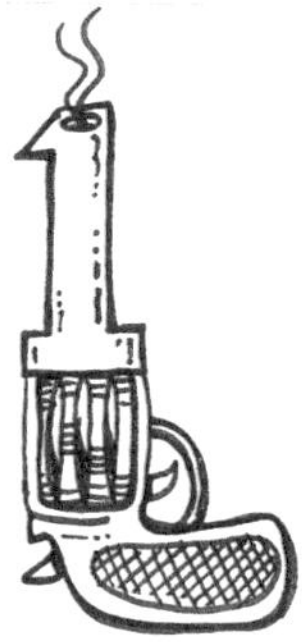

1. GUN

2. SHOE

3. TREE

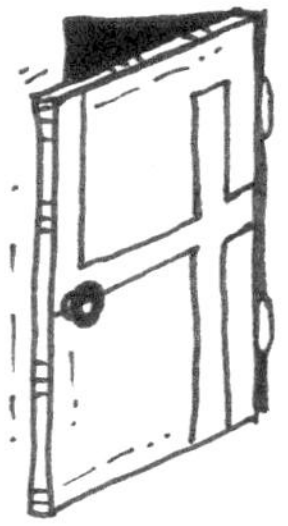

4. DOOR

5. HIVE

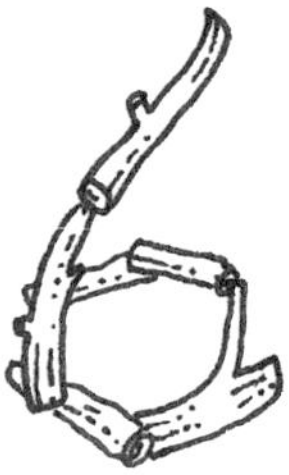

6. STICKS

7. HEAVEN

8. GATE

9. SIGN

10. HEN

APPENDIX D
A BETTER MEMORY AT ANY AGE

Americans are frightened of losing their memory! Half of men and women over 42, in a national survey said that it was their most significant fear.

GUESS WHAT!!! Scientists now know that memory loss is not part of normal aging! You can live into your seventies, eighties, and even nineties, as sharp as a tack. You don't have to accept that memory loss is going to happen to you. A large study of about 6,000 older people published in the Journal of the American Medical Association in 1999 found that 70% of the subjects had no decline in memory or thinking during the ten year study.

Dr. Irwin Rosenberg, head of the Nutrition and Neurocognition Laboratory at Tufts University in Boston, (2007) said, "Cognitive decline is not necessarily a feature of normal aging. But aging may affect the brain's need for nutrients."

"Adjusting dietary intake of certain nutrients could eventually prove to be the key to forestalling cognitive decline." As we age, our need for brain nutrients increases. If we don't supply our brain with these essential nutrients, then we are much more likely to have memory and other learning problems.

Scientific studies show that changes in nutrition and lifestyle may reduce your risk of memory loss by as much as 70%. The things you can do to reverse memory loss are:

1. Get really serious about the possible side effects from your medications. Some of the culprits might be:

   Chronic use of aspirin- Advil and Ibuprofen

   Valium, antidepressants, sedatives

   Xanax, Lomotil, Benadryl, Elavil, Ambien and other sleeping Pills.

   This is according to Dr. Ali Tabatabai, an internist in Laguna Woods, California and Public Citizens Health Research Group.

2. One of the most significant scientific breakthroughs about memory is that stress shrinks your brain. Stress produces a hormone, cortisol, which adversely affects the part of the brain that learns

and remembers, the Hippocampus. Too much cortisol blocks the hippocampus from making new cells and thus the hippocampus shrinks. Having a larger volume of brain cells in the hippocampus compared to a smaller number may help prevent memory loss as you age. (Yale University, Science Daily 2008) *Therefore it is imperative that you find a way to reduce your stress. You must learn to control negative emotions such as anxiety, anger and depression. Everyday pressure is normal but you CAN LEARN TO CONTROL these emotions. It's an inside job! You must also learn to relax and avoid these negative emotions at night because it is crucial that you have a relaxed seven hours of restful sleep.*

3. You must also have a balanced diet with plenty of fresh fruits and vegetables. Eat every three hours and never skip a meal, especially breakfast. Cortisol rises when your blood sugar goes down too low. Every time you eat, include a protein (almonds, walnuts, chicken, fish) a good carbohydrate (carrots, a banana, a piece of whole grain bread) and a good fat (peanut butter, nut butter, olive oil)

4. Eliminate sodas. A sudden intake of sugar spikes your blood sugar and consequently increases your cortisol. Both can be damaging to the blood vessels and cells in your brain. If you do just one thing – eliminate all sodas (including diet) and sweetened beverages. (American Journal of Public Health. 2007)

5. Exercise reduces stress cortisol. Tai Chi, yoga, swimming and Dancing have all been shown to reduce cortisol. Exercise plus mental stimulation (crossword puzzles) significantly improve memory and cognitive performance. (Mayo Clinic, 2007)

6. Supplements to improve memory include:

Soy-based supplement
Omega-3 fish oil
Antioxidants, Vitamin E, C, and blueberries
(New England Journal of Medicine, 1997)
Complete Vitamin B formula that includes choline
(Iowa State University, 2007)
Vitamin D3, 1,000-2,000 IU each day

Taken from Mind Matters Vol. 1

PHYSICAL IMPROVEMENT OF YOUR MEMORY EXERCISE CREATES NEW BRAIN CELLS

"Aerobic exercise increases the supply of a protein called brain-derived neurotrophic factor, or BDNF, which protects brain neurons and promotes the growth of new nerve cells and synapses that are related to learning and memory," says Charles Hillman, director of the Neurocognitive Kinesiology Laboratory at the at the University of Illinois at Urbana- Champaign.

BDNF is active in the "hippocampus," an area deep in the brain that is essential for memory, especially for what is called relational memory. That's the ability to make logical connections among pieces of information and to navigate in space. The "hippocampus" tends to shrink as people age, and it's one of the first parts of the brain to be damaged in Alzheimer's disease.

"Impaired spatial -memory-having trouble finding your way or knowing what goes where-is a major reason older people lose their independence," points out one other Illinois researchers, Art Kramer. *"This study shows that there is a significant and substantial relationship between how fit you are and how good the kind of memory is, that we need all of the time."*

Aerobic exercise also appears to boost executive function. That's the ability to plan or make decisions, correct errors, or react to new situations.

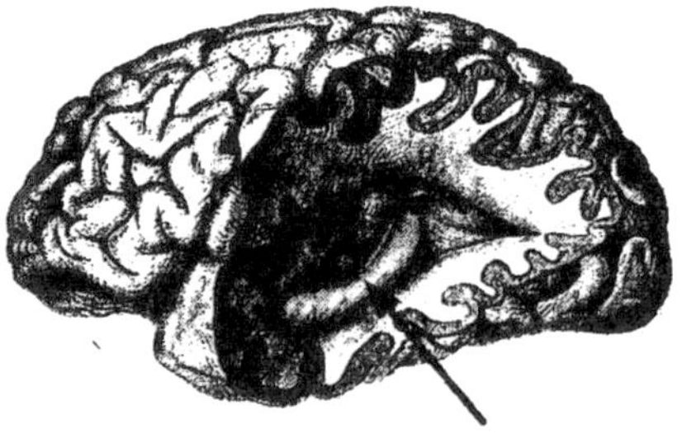

Aerobic exercise may increase the size of the hippocampus, an area deep In the brain that's critical for memory.

HIPPOCAMPUS

APPENDIX E

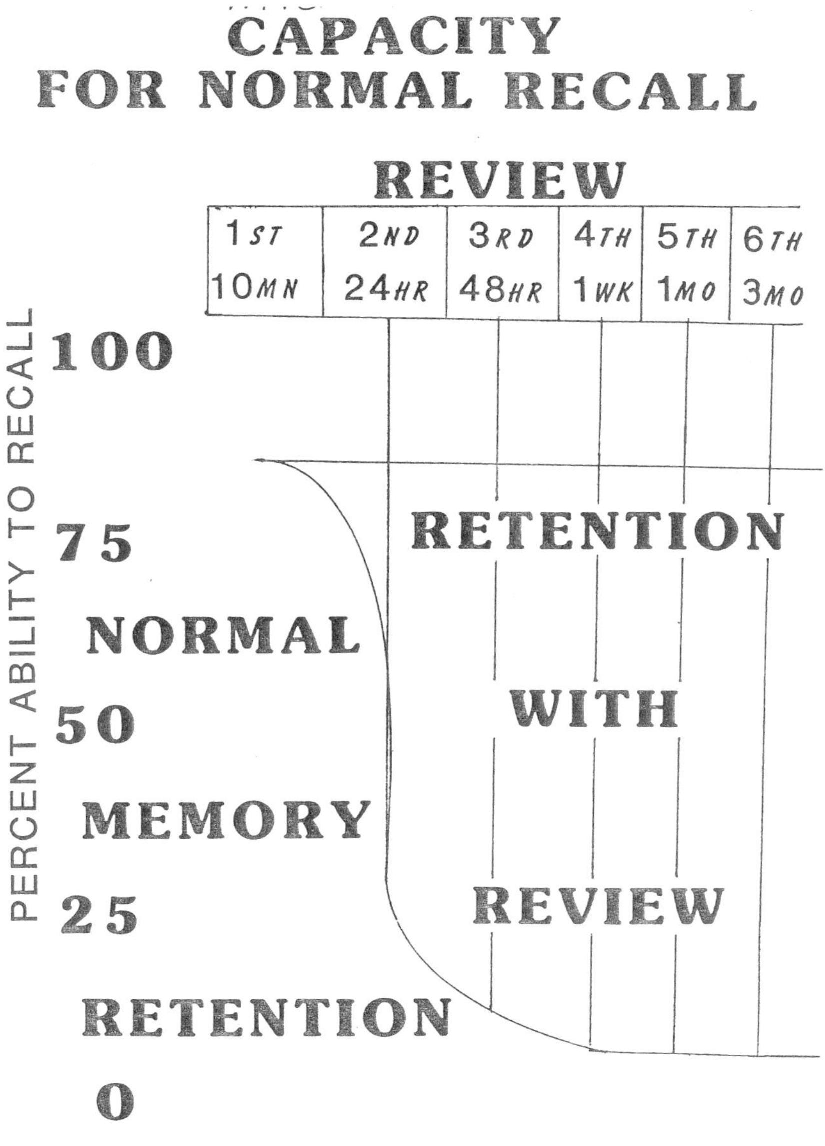
CAPACITY
FOR NORMAL RECALL
REVIEW
1ST
10MN
2ND
24HR
3RD
48HR
4TH
1WK
5TH
1MO
6TH
3MO
PERCENT ABILITY TO RECALL
100
75
50
25
0
NORMAL
MEMORY
RETENTION
RETENTION
WITH
REVIEW

Made in the USA
Monee, IL
05 November 2026

67385920R00098